Advances in Sustainability Science and Technology

The book series aims at bringing together valuable and novel scientific contributions that address the critical issues of renewable energy, sustainable building, sustainable manufacturing, and other sustainability science and technology topics that have an impact in this diverse and fast-changing research community in academia and industry.

The areas to be covered are

- Climate change and mitigation, atmospheric carbon reduction, global warming
- Sustainability science, sustainability technologies
- Sustainable building technologies
- Intelligent buildings
- Sustainable energy generation
- Combined heat and power and district heating systems
- Control and optimization of renewable energy systems
- Smart grids and micro grids, local energy markets
- Smart cities, smart buildings, smart districts, smart countryside
- Energy and environmental assessment in buildings and cities
- Sustainable design, innovation and services
- Sustainable manufacturing processes and technology
- Sustainable manufacturing systems and enterprises
- Decision support for sustainability
- Micro/nanomachining, microelectromechanical machines (MEMS)
- Sustainable transport, smart vehicles and smart roads
- Information technology and artificial intelligence applied to sustainability
- Big data and data analytics applied to sustainability
- Sustainable food production, sustainable horticulture and agriculture
- Sustainability of air, water and other natural resources
- Sustainability policy, shaping the future, the triple bottom line, the circular economy

High quality content is an essential feature for all book proposals accepted for the series. It is expected that editors of all accepted volumes will ensure that contributions are subjected to an appropriate level of reviewing process and adhere to KES quality principles.

The series will include monographs, edited volumes, and selected proceedings.

More information about this series at https://link.springer.com/bookseries/16477

Rajib Kumar Bhattacharjya · Bipul Talukdar ·
Konstantinos L. Katsifarakis
Editors

Sustainable Water Resources Management

Proceedings of SWARM 2020

Springer

Editors
Rajib Kumar Bhattacharjya
Department of Civil Engineering
Indian Institute of Technology Guwahati
Guwahati, India

Bipul Talukdar
Department of Civil Engineering
Assam Engineering College
Guwahati, India

Konstantinos L. Katsifarakis
School of Civil Engineering
Aristotle University of Thessaloniki
Thessaloniki, Greece

ISSN 2662-6829 ISSN 2662-6837 (electronic)
Advances in Sustainability Science and Technology
ISBN 978-981-16-7534-8 ISBN 978-981-16-7535-5 (eBook)
https://doi.org/10.1007/978-981-16-7535-5

This Springer imprint is published by the registered company Springer Nature Singapore Pte Ltd.
The registered company address is: 152 Beach Road, #21-01/04 Gateway East, Singapore 189721, Singapore

Preface

The rapid growth of population on the earth's surface is forcing industrialization and urbanization. In developing countries, urbanization is more often unplanned and disorganized, resulting in various hazards due to ecological imbalance. Rapid unplanned urbanization has converted many forest areas into agricultural or urban sectors and filled up low-lying regions to meet the resource need to support the increased population. Unprecedented natural calamities, like devastating floods, riverbank erosion, surface erosion, and landslides, can be regarded as a manifestation of nature's reaction to restore balance. The increased emission of greenhouse gases due to industrialization, long-distance transport of goods and raw materials and the continuation of unscientific agricultural practices in many parts of the world has led to climate change. Accelerated climate change is expected to have a major impact on the future sustainability of the ecosystems.

On the other hand, the increase in population has reduced the per capita availability of water, resulting in unplanned exploitation of both surface water and groundwater. Due to overexploitation, groundwater has been depleted in many parts of the world. As a result, it becomes uneconomical to withdraw groundwater from aquifers. The overexploitation of groundwater as well as increased number of pollution sources may also contaminate the water resources. Therefore, scientific studies on sustainable water management under changing climate are vital for the appropriate management of the world's water resources.

This book brings high-quality selected research articles from the international conference on Sustainable Water Resources Management, SWARM2020, jointly organized by Assam Engineering College Guwahati and Indian Institute of Technology Guwahati, on water management and planning, urban water management, climate change and global warming, management of groundwater and aquifer remediation, water conservation, water quality and pollution control, management of transboundary rivers, advanced hydrological modeling and hydro-disaster risk management under one umbrella.

The subject matter of the book is divided into four parts. Part I deals with the impact of climate change on water. Several pieces of research conducted in different parts of the world reveal that the impact of climate change is quite visible now and will be even more severe in the future. The recent IPCC report published under the title 'Climate Change 2021: The Physical Science Basis' stated that there would be more intense and frequent heatwaves and humid heat stress during the twenty-first century over South Asia. There are six chapters in this part that discussed different aspects of climate change on water resources.

Hydrologic and hydraulic modeling are important components of water resource planning and management. Part II, comprising seven chapters, deals with the modeling aspect of water resources. The first three chapters deal with hydrologic modeling using SWAT. The hydraulic modeling of river flow has been presented in the next three chapters of this part. A case study on the optimization of an irrigation canal has also been presented in this part.

Most civilizations have flourished on the banks of a river, and thus, the river is considered the lifeline of every civilization. Though initially, people did not live on the flood plain, over the year, they occupied the fertile land on the flood plain to have a better livelihood and started staying on the flood plain itself. As a result, many densely populated areas are now suffering from flood and erosion problems. Part III deals with the various aspects of river management. There are nine chapters in this part that deal with the multiple aspects of river management.

As reported by many researchers, surface and groundwater have been contaminated due to various human activities and some natural phenomena. Part IV, comprising two chapters, deals with the water quality aspect and management of water resources.

We hope that this book can serve as a reference material for the research and teaching fraternity working on water resources and management.

Guwahati, India — Rajib Kumar Bhattacharjya
Guwahati, India — Bipul Talukdar
Thessaloniki, Greece — Konstantinos L. Katsifarakis

Contents

Editors and Contributors

About the Editors

Prof. Rajib Kumar Bhattacharjya is a professor in the Department of Civil Engineering, Indian Institute of Technology Guwahati, India. He received his Bachelor's and Master's Degree in Civil Engineering from Gauhati University, India in the year 1993 and 1995, respectively and Ph.D. in Civil Engineering from Indian Institute of Technology Kanpur, India in the year 2004. His current research interests include computational hydraulics, the impact of climate change, assessment and management of groundwater resources, flood modeling, optimization methods, and artificial neural networks. He has more than 25 years of teaching and research experience and has authored more than 100 peer-reviewed scientific publications in various reputed international journals and conference proceedings. He has jointly edited three books *Urban Hydrology, Watershed Management, and Socio-Economic Aspects*, *Urban Ecology, Water Quality and Climate Change*, and *Nature-Inspired Methods for Metaheuristics Optimization: Algorithms and Applications in Science and Engineering*, published by Springer. As a faculty at Indian Institute of Technology Guwahati, he has taught courses on subsurface hydrology, surface water hydrology, hydraulics, water resources planning and management, optimization methods, *etc.* He has also been a visiting professor at other institutes, including the Dalhousie University, Halifax, Canada, and Ecole Centrale Nantes, France. More about his research and academic activities can be found at http://www.iitg.ac.in/rkbc

Prof. Bipul Talukdar is currently working as a Professor in the Department of Civil Engineering, Assam Engineering College, Guwahati. He received his Bachelor's and Master's Degree in Civil Engineering from Dibrugarh University and Gauhati University, India in the year 1988 and 1990, respectively and Ph.D. in Hydrology from Indian Institute of Technology Roorkee, India in the year 2000. He has been in teaching and research for the last 28 years in the areas of water resources systems, flood and erosion management of large river systems, watershed management, remote sensing and GIS, restoration and management of lakes and rivers and has authored more than

50 peer-reviewed scientific publications in journals and conference proceedings. He has guided 35 Masters and 4 research students till now and also worked as a guest editor for the special publication *Sustainable Development—A Civil Engineering Perspective* by Springer Nature. He has been with various technical committees of government related to water resources management and was an academic visitor to the University of Melbourne, Australia. Details about him can be found at http://civil.aec.ac.in/people/faculty

Prof. Konstantinos L. Katsifarakis is a professor of the Civil Engineering Department in Aristotle University of Thessaloniki, Greece. He holds a Diploma in Civil Engineering, a Bachelor in Geology, an M.S. Degree in Engineering Mechanics and a Doctoral Degree in Civil Engineering. His research interests include groundwater flows and contamination control, water resources management, optimization techniques, low enthalpy geothermal energy, environmental impact mitigation, solid waste management, education of engineers and history of hydrology. He has authored more than 100 papers published in peer-reviewed scientific journals and conference proceedings, and he has served as editor of 2 books and 7 proceedings of international conferences. He has taught many undergraduate and graduate courses and has supervised more than 230 undergraduate and graduate Diploma Theses and 10 Ph.D. Theses. He has served Aristotle University of Thessaloniki as Coordinator of the Environment Council, Director of a Graduate Studies Program, member of the University Board and Dean of the Faculty of Engineering.

Contributors

E. K. Abdul Rahman Faculty of Engineering, Universiti Teknologi Brunei, Gadong, BE, Brunei Darussalam

Pathak A. Abhishek Department of Civil Engineering, Dayananda Sagar College of Engineering, Bangalore, Karnataka, India

S. N. Amirah Arun Faculty of Engineering, Universiti Teknologi Brunei, Gadong, BE, Brunei Darussalam

Swapnali Barman Center for Flood Management Studies, National Institute of Hydrology, Dispur, Guwahati, India

Anupal Baruah Department of Civil Engineering, Indian Institute of Technology, Guwahati, India

Rajib Kumar Bhattacharjya Department of Civil Engineering, Indian Institute of Technology Guwahati, Guwahati, India

Basudev Biswal Department of Civil Engineering, IIT Bombay, Mumbai, India; The Interdisciplinary Programme (IDP) in Climate Studies, IIT Bombay, Mumbai, India

Leena Chetia Department of Architecture and Regional Planning, Indian Institute of Technology Kharagpur, Kharagpur, West Bengal, India

Krishna Kamal Das Civil Engineering Department, Bongaigaon Polytechnic, Bongaigaon, Assam, India

Priyam Deka Department of Civil Engineering, Indian Institute of Technology, Guwahati, India

Ranjit Deka NEHARI, Brahmaputra Board, Guwahati, India;
IIT, Guwahati, India

Dipsikha Devi Department of Civil Engineering, Indian Institute of Technology Guwahati, Guwahati, Assam, India

Thiyam Tamphasana Devi Department of Civil Engineering, National Institute of Technology, Manipur, Imphal, Manipur, India

Richa Dhawale Department of Architecture and Regional Planning, Indian Institute of Technology Kharagpur, Kharagpur, West Bengal, India

Monisha Dutta Civil Engineering Department, Jorhat Engineering College, Jorhat, India

Mrinal Kumar Dutta Civil Engineering Department, Jorhat Engineering College, Jorhat, India

Pulendra Dutta Department of Civil Engineering, Jorhat Engineering College, Jorhat, India

Arghya Ghosh Department of Civil Engineering, Indian Institute of Technology Guwahati, Guwahati, Assam, India

Sanjib Gohain Department of Civil Engineering, Assam Engineering College, Guwahati, India

Jayshree Hazarika Civil Engineering Department, Assam Engineering College, Guwahati, Assam, India

Nayana Merin Joy Department of Architecture and Regional Planning, Indian Institute of Technology Kharagpur, Kharagpur, West Bengal, India

Snigdha Kalita Civil Engineering Department, Jorhat Engineering College, Jorhat, India

Suresh A. Kartha Department of Civil Engineering, Indian Institute of Technology Guwahati, Guwahati, Assam, India

P. K. Khaund Civil Engineering Department, Jorhat Engineering College, Jorhat, India

Sandip Mondal Department of Earth and Environmental Studies, National Institute of Technology Durgapur, Durgapur, West Bengal, India

Upadhyay Mudita Coca-Cola Department of Regional Water Studies, TERI School of Advanced Studies, Vasant Kunj, New Delhi, India

Debbarma Nilotpal Civil Engineering Department, NIT Agartala, Tripura, India

Roy Parthajit Civil Engineering Department, NIT Silchar, Assam, India

Choudhury Parthsarathi Civil Engineering Department, NIT Silchar, Assam, India

Nitesh Patidar Ground Water Hydrology Division, NIH, Roorkee, India

Saikat Kumar Paul Department of Architecture and Regional Planning, Indian Institute of Technology Kharagpur, Kharagpur, West Bengal, India

Sreeja Pekkat Department of Civil Engineering, Indian Institute of Technology Guwahati, Guwahati, Assam, India

Sultana Hadia Rahman Department of Civil Engineering, Assam Engineering College, Guwahati, Assam, India

Waikhom Rahul Singh Center for Flood Management Studies, Guwahati, National Institute of Hydrology, Dispur, India

U. Ratnayake Faculty of Engineering, Universiti Teknologi Brunei, Gadong, BE, Brunei Darussalam

Arup Kumar Sarma Department of Civil Engineering, Indian Institute of Technology Guwahati, Guwahati, Assam, India

Bibhash Sarma Department of Civil Engineering, Assam Engineering College, Guwahati, Assam, India

Dipima Sarma Climate Change Cell, Forest Department, Government of Assam, Guwahati, Assam, India

Dhruba Jyoti Sarmah Department of Civil Engineering, Indian Institute of Technology Guwahati, Guwahati, India

Baldev Setia Department of Civil Engineering, NIT Kurukshetra, Kurukshetra, Haryana, India

S. Shams Faculty of Engineering, Universiti Teknologi Brunei, Gadong, BE, Brunei Darussalam

Durga Sharma Department of Civil Engineering, IIT Hyderabad, Sangareddy, India

Smita Sharma Department of Economics, Panjab University, Chandigarh, India

Anupama Shashni Department of Economics, Panjab University, Chandigarh, India

Alemayehu A. Shawul Assistant Professor, Natural Resource Management Department, Madda Walabu University, Bale Robe, Ethiopia

N. C. Sanjay Shekar Department of Civil Engineering, JSS Academy of Technical Education, Bangalore, Karnataka, India

M. A. Sherly Coca-Cola Department of Regional Water Studies, TERI School of Advanced Studies, Vasant Kunj, New Delhi, India

Agarwal Shivam Civil Engineering Department, NIT Silchar, Assam, India

Nameirakpam Momo Singh Department of Civil Engineering, National Institute of Technology, Manipur, Imphal, Manipur, India

Chakma Sumedha Civil Engineering Department, Indian Institute of Technology Delhi, Hauz Khas, New Delhi, India

Amrutha Suresh Department of Civil Engineering, Indian Institute of Technology Guwahati, Guwahati, Assam, India

Annu Taggu Department of Agricultural Engineering, NERIST, Nirjuli, India

Gaurav Talukdar Department of Civil Engineering, Indian Institute of Technology, Guwahati, India

Jaivir Tyagi National Institute of Hydrology, Roorkee, Roorkee, India

Rahul Singh Waikhom Center for Flood Management Studies, NIH, Guwahati, India

Thronlem Winkangshu Department of Civil Engineering, National Institute of Technology, Manipur, Imphal, Manipur, India

Lalit Yadav Department of Civil Engineering, NIT Kurukshetra, Kurukshetra, Haryana, India

Climate Change

Chapter 1
Climate Change and Its Impact on Surface Runoff Characteristics of an Urban Catchment

Amrutha Suresh(✉) and Sreeja Pekkat

Department of Civil Engineering, Indian Institute of Technology Guwahati, Guwahati, Assam 781039, India
{a.suresh,sreeja}@iitg.ac.in

Abstract. Flooding can occur naturally due to heavy rain, exceptionally high tides, snowmelt, or human-made causes such as dam break, breach of river embankment, and so on. Urban flooding and its associated fury have become very common in the recent past. The combined effects of urbanization and climate change have increased the risk of flooding in cities that can be attributed to various factors. The urbanization causes imperviousness of the surface due to more built-up areas, which lead to reduced infiltration, thus, an increment in the surface runoff. Climate change has affected the quantity, intensity, and frequency of precipitation, which made the urban catchments more vulnerable to flooding. The present study aims to investigate the potential impacts of climate change on the surface runoff characteristics of an urbanized watershed. The future climate scenario corresponding to the study area was obtained from global climate models (GCM) that has shown superior performance in forecasting the future climate. As the GCM outputs are for a larger grid, they must be downscaled to the local/regional scale to obtain the corresponding climatic variables. In this study, we have used the delta change method of downscaling. Further, the runoff was simulated using the stormwater management model (SWMM), considering the future scenario. The results from precipitation analysis indicate an increase in precipitation in the future. Consequently, the simulated runoff has shown an increment in the number of flood events in 2025, 2050, 2075, and 2100. The present study has also found that the continuous rainfall simulation was resulting in an underestimation of the urban runoff. Therefore, for urban applications, event-based runoff simulation would be a better choice for characterizing the future urban flood events.

Keywords: Climate change · GCMs · Urban runoff · Urbanization

1 Introduction

Flooding can occur naturally due to heavy rain, exceptionally high tides, snowmelt, or human-made causes such as dam break, breach of river embankment, and so on. Flooding is conceived as a natural disaster that leads to loss of life, damage to property, and wreak havoc on agriculture and infrastructure [8] every year. As the occurrence of a

R. K. Bhattacharjya et al. (eds.), *Sustainable Water Resources Management*,
Advances in Sustainability Science and Technology,
https://doi.org/10.1007/978-981-16-7535-5_1

flood depends on the climate and the surface characteristics, it is apparent that the flood characteristics are not unique across the world. For countries like India, where about 80% of the annual rainfall receives during the southwest monsoon (National disaster management guidelines management of flood report, 2008), the occurrence of the flood would be more frequent. Expectedly, the high-intensity, short-duration rainfall causes the river to overflow and lead to flooding in the nearby areas. As per the Indian flood management report (2015), around 40 million hectares (M ha) of land out of the total area are prone to flooding.

Urban flooding and its associated fury have become very common in this century, and it is visible from various flood events like Hyderabad (2008), Kolkata (2007), Chennai (2004 and 2015), Mumbai (2005), Delhi (1924, 1947, 1976, 1978, 1988, and 1995), Noida (2010), and Guwahati (2010) [6]. Rapid urbanization, high-intensity short-duration rainfall, and climate change are the most important causes of urban flooding. Studies have shown that climate change has increased the occurrence of flash floods both in urban and rural areas [7].

Nevertheless, one of the leading causes of urban flooding is climate change, which is happening day by day in a haphazard manner. As per the IPCC AR5 report, the energy balance of the earth is disturbed because of the change in the concentration of greenhouse gases (GHGs), aerosols, land cover, and solar radiation. The anthropogenic activities are considered to be one of the leading causes of the increase in GHGs concentration [5]. The climate change is evident from the changing patterns of rainfall, rise in temperature, sea-level rise, melting of ice, and increase in extreme events. Some of the adverse effects of climate change are altered precipitation amounts, intensities, and frequencies, which increased the vulnerability of flooding in urban areas [15]. To fully understand the havoc of flooding and to develop proper flood management strategies, a better knowledge of changes in climate and its influence on the flood occurrences need to be studied.

The present study aims to investigate the potential impact of climate change on the urban runoff characteristics for an urbanized watershed in the northeastern part of India.

2 Study Area and Data

The study area, the Urban Guwahati, in the state of Assam, India, is located on the bank of river Brahmaputra. The geographic location of Guwahati is between 26° 4′ 45″ N and 26° 13′ 25″ North Latitude and between 91° 34′ 25″ and 91° 52′ 00″ East Longitude. Urban Guwahati is one of the largest commercial and fast-growing cities in Northeastern India. The city is a part of the Kamrup district of Assam state and located on the undulating plain of a varying altitude of 49.5–55.5 m above the mean sea level (MSL). Guwahati has a subtropical and humid climate with an average annual rainfall of 1729.5 mm. The Guwahati city is divided into seven watersheds. The most flood-prone region has a catchment area of 28.3 km^2, which is the Bharalu watershed. The Bharalu watershed was considered to demonstrate the methodology. The watershed includes residential, commercial, institutional, and business areas, and the river Bharalu carries the entire stormwater of the basin and discharges it to the river Brahmaputra.

The daily rainfall data for the following years 2006, 2011, and 2016 were collected from the Regional Meteorological Centre (RMC), Guwahati. Future rainfall was derived from the Coupled Model Intercomparison Project 5 (CMIP5) GCM. In this study, the MPI-ESM-LR model developed by the Max Planck Institute for Meteorology (MPI-M) was used for downscaling the precipitation. Representative concentration pathway (RCP) 4.5 was considered. The period from 1980 to 2005 was considered as a control period, and downscaling was done for 2006 to 2100. Drainage network map was collected from the Guwahati Municipal Development Authority (GMDA) and Guwahati East Water Resources Division. The satellite image for the year 2011 was acquired from National Remote Sensing Agency (NRSA) in India. Topo sheet and base map of Guwahati city were collected from the Center of Excellence (COE) project, Water Resources Engineering and Management, IIT Guwahati, and GMDA, respectively.

3 Methodology

Downscaling

Climate changes are well captured in global climate models (GCMs). However, the performance of GCMs is only better at coarser scales ($>10^4$ km^2) and poorly at the local/regional scales [3]. Thus, the climatic variables corresponding to the larger grid, obtained from GCMs, need to be downscaled to local/regional scale. The present study uses the delta change method to downscale the precipitation. Delta change method is one of the most popular statistical downscaling methods. Several types of delta change methods available in literature can be categorized considering the temporal scale, temporal resolution, mathematical formulation [1], or simple delta change to advanced delta change method [13]. In the delta change method, the future climate is projected based on a change factor, which is a ratio between the future and historical run of the same GCM. This factor is then applied to the observed time series to transform it into future series. The change factor is calculated monthly to take the effect of seasonality. The underlying assumption involved in this method is that the bias in the current climate would remain the same in future climate, and the change factor calculated is independent of the spatial scale [14]. In this study, an advanced delta change method called the quantile perturbation method is applied. The extreme monthly value of the control and future precipitation is calculated from the GCM. Extreme rainfall is defined in this study as precipitation with a return period (T) higher than one year. The return period is calculated empirically based on the rank of precipitation values (n/i, where n and i are the length of the study period and rank, respectively; $i = 1$ for the highest value) [13]. The change factor for precipitation is calculated as the ratio of GCM future period to that of GCM control period. The estimated change factor is then applied to the observed precipitation series to generate future values.

Imperviousness Determination

The rapid and unplanned urbanization has modified the watershed characteristics by making the surface impervious, thereby reducing infiltration and increasing the surface runoff. Imperviousness is mainly measured in terms of the total impervious area (TIA) and the effective impervious area (EIA). TIA gives the area that does not allow water to infiltrate, whereas EIA is a portion of TIA which is hydraulically connected to the drainage system. EIA was found to be a better predictor of runoff, whereas TIA causes overestimation of the runoff [2, 4, 9, 11]. However, direct computation of EIA is challenging as it requires high-resolution satellite imageries and field visits to ensure the collection system's connectivity. [12] had developed an empirical equation for the study area to determine the EIA from TIA as given by Eq. (1). The sensitivity coefficient of Eq. (1) was found to be 59%. Sensitivity coefficient shows how the change in one variable affects the other variable.

$$\text{EIA} = 0.0031 * (\text{TIA})^{2.21} \quad (1)$$

In this study, we had used this relationship to estimate the EIA for the study area.

Runoff Modeling

The runoff modeling was carried out by using the stormwater management model (SWMM) as it is applicable in hydrologic modeling of an urban catchment [10]. SWMM is a dynamic rainfall–runoff model developed by the Environmental Protection Agency (EPA) in the USA. SWMM is used for continuous or event-based simulation of the runoff quantity and quality.

SWMM consists of two components: (i) hydrological component and (ii) the hydraulic component. In the hydrological module, the watershed is modeled as a non-linear reservoir with rainfall as input to generate the runoff. Surface runoff occurs when the depth of water exceeds the maximum depression storage. The infiltration occurring in the previous regions can be modeled in SWMM by three different methods, such as Horton, Green–Ampt, and Curve number. In this study, we had used the Green–Ampt method to find the infiltration loss of catchments. The runoff generated by the hydrological module of SWMM was transported to the hydraulic module which models the drainage network based on the Saint Venant's equations. SWMM can perform three different types of hydraulic routing: (i) steady flow routing, (ii) kinematic wave routing, and (iii) dynamic wave routing. In this study, we have considered the dynamic wave routing, which solves the complete one-dimensional Saint Venant's equations of flow for the entire drainage networks.

4 Results and Discussion

Downscaling

The change factor is calculated monthly to take the effect of seasonality. Figure 2 shows the monthly precipitation change factor. In this study, we have divided the downscaled rainfall into three periods, i.e., near future (2020–2031), middle of the century (2032–2057), and end of the century (2058–2100). For the near future, the precipitation change factor for the April month was high, which indicate an increase in the rainfall. For the middle of the century, the December month was showing the highest change factor of about 1.5, whereas for the end of the century, the change factor for May, June, and July was showing higher value.

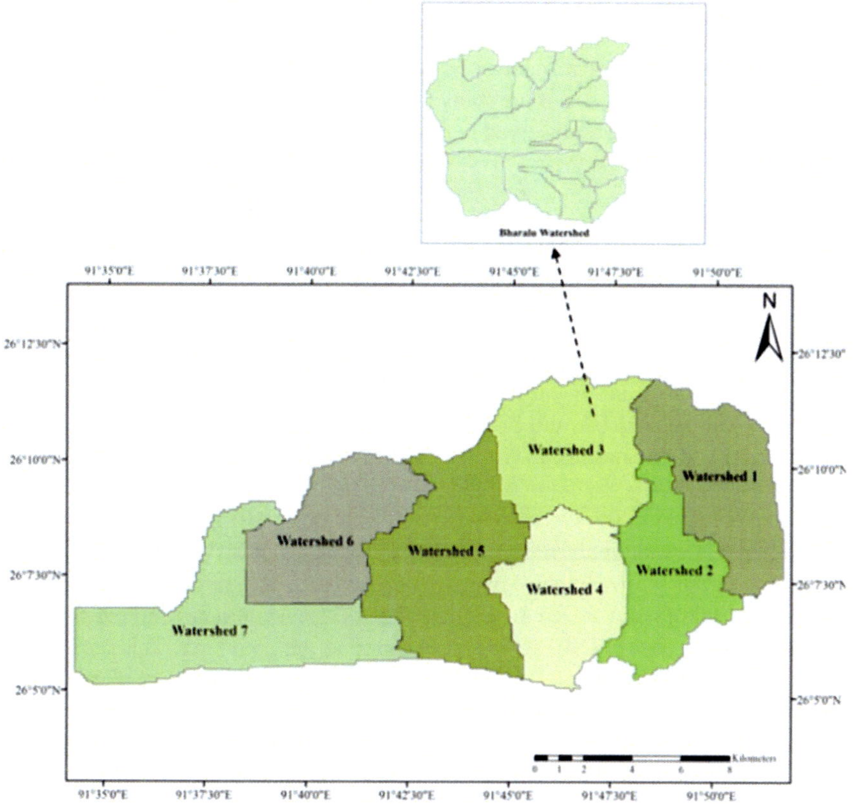

Fig.1 Location of Bharalu watershed of Guwahati city

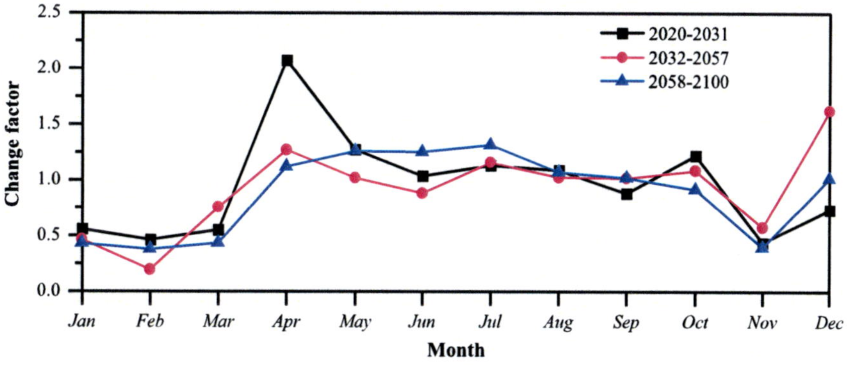

Fig. 2 Monthly precipitation change factor

Imperviousness Determination

The land use analysis shows that the percentage of the built area was 36.4% in 2011. The results of the image analysis of the study area show that in the year 2011, the percentage of EIA was 12.22%. It was observed that the study area had a maximum imperviousness of 33.57%.

Runoff Modeling

In this study, we had analyzed the impact of climate change on runoff by keeping all other physical characteristics of the basin constant and assuming land use corresponding to the year 2011 for future years. Figure 3 shows the runoff simulated for three historical continuous rainfall data for the years 2006, 2011, and 2016 and four future rainfall for the years 2025, 2050, 2075, and 2100. The number of flood events was found to increase in 2025, 2050, 2075, and 2100. The peak runoff was 3.87 m^3/s, 6.771 m^3/s, 4.573m^3/s, 3.904m^3/s, 5.337m^3/s, 5.060m^3/s, and 4.615m^3/s for 2006, 2011, 2016, 2025, 2050, 2075, and 2100 rainfall, respectively. Runoff simulation using continuous rainfall has led to an underestimation of runoff.

To confirm the above hypothesis, we have used a 15-min rainfall event for modeling the runoff. Figure 4 shows the runoff simulated for a 15-min rainfall event. The peak simulated runoff was 164.307 m^3/s and occurred at the 26th hour. This result confirms that continuous rainfall simulation leads to an underestimation of runoff. Thus, for urban flood studies, an event-based simulation must be used to get a better understanding of the flood characteristics.

Table 1 shows that the peak runoff based on the continuous rainfall data has been under-predicted by 0.0269 and 0.0588 times as compared to runoff obtained by considering rainfall events for the years 2006 and 2011, respectively.

5 Conclusion

The influence of climate change in terms of rainfall on surface runoff was studied for an urban catchment. The delta change method of downscaling was used for projecting future precipitation. The analyses show an increase in rainfall in the future. The runoff was simulated using a dynamic rainfall–runoff model, the SWMM. Intensification in the number of flood events was observed in the future for Guwahati. The results indicate that the use of continuous rainfall has led to an underestimation of the runoff when compared to event-based runoff. Therefore, from the flood management perspective, such an underestimation is objectionable for the urban catchment. Hence, it is recommended to use event rainfall with realistic peak rainfall intensity for runoff determination.

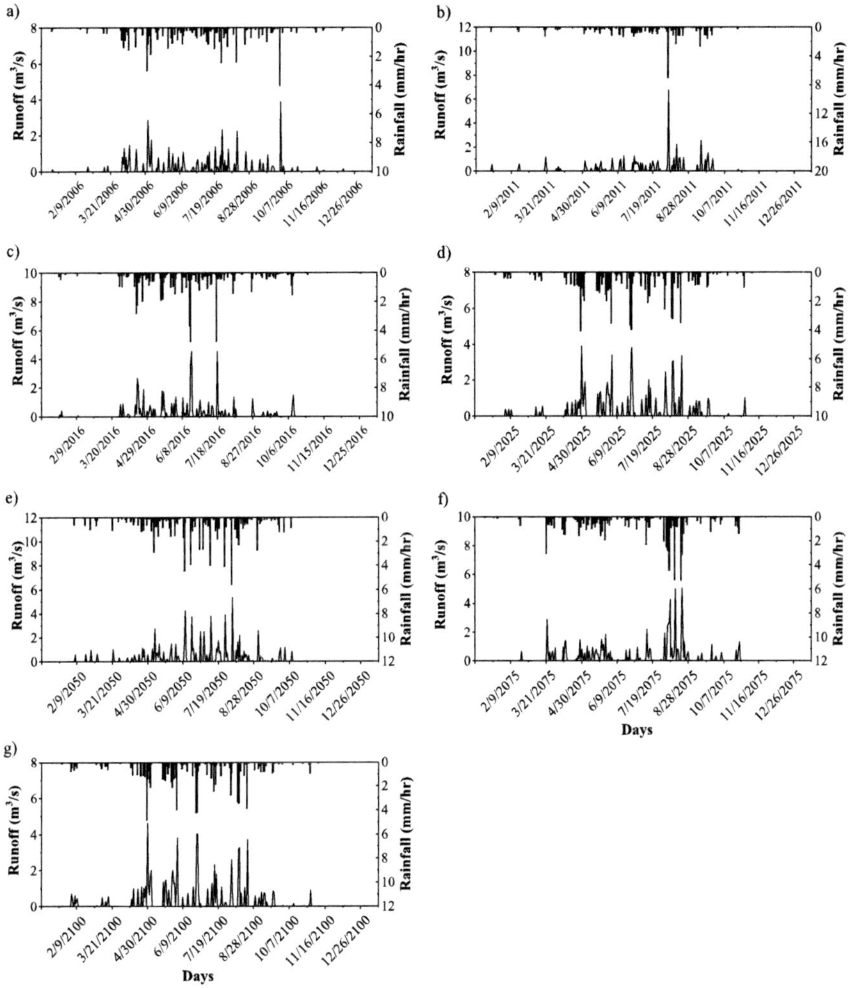

Fig. 3 Simulated runoff using continuous rainfall data of **a** 2006, **b** 2011, **c** 2016, **d** 2025, **e** 2050, **f** 2075, **g** 2100 rainfall

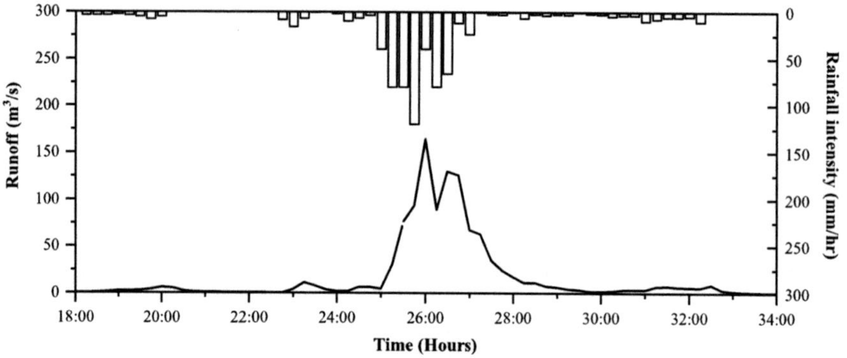

Fig. 4 Runoff simulated for event-based rainfall

Table 1 Underestimation of peak runoff using continuous rainfall

	Peak runoff (m^3/s)		
Year	Continuous rainfall (A)	Rainfall Event (B)	Under prediction (A/B)
2006	3.87	143.819	0.0269
2011	6.771	115.189	0.0588

References

1. Anandhi A, Frei A, Pierson DC, Schneiderman EM, Zion MS, Lounsbury D, Matonse AH (2011) Examination of change factor methodologies for climate change impact assessment. Water Resour Res 47(3)
2. Ebrahimian A, Gulliver JS, Wilson BN (2016) Effective impervious area for runoff in urban watersheds. Hydrol Process 30(20):3717–3729
3. Ghosh S, Mujumdar PP (2008) Statistical downscaling of GCM simulations to streamflow using relevance vector machine. Adv Water Resour 31(1):132–146
4. Han WS, Burian SJ (2009) Determining effective impervious area for urban hydrologic modeling. J Hydrol Eng 14(2):111–120
5. Nicholls R, Hoozemans F, Marchand M (1999) Increasing flood risk and wetland losses due to global sea-level rise: regional and global analyses. Global Environ Change 9(SUPPL.):S69–S87
6. Patil A (2015) Urban hydrology, need of India. Environ We Int J Sci Technol 10:29–36
7. Perumal M, Sahoo B (2010) Real-time flood forecasting by a hydrometric data-based technique. Natural and Anthropogenic Disasters: Vulnerability, Preparedness, and Mitigation
8. Rao KHVD, Rao VV, Dadhwal VK, Behera G, Sharma JR (2011) A distributed model for real-time flood forecasting in the Godavari Basin using space inputs. Int J Disaster Risk Sci 2(3):31–40
9. Ravagnani F, Pellegrinelli A, Franchini M (2009) Estimation of urban impervious fraction from satellite images and its impact on peak discharge entering a storm sewer system. Water Resour Manage 23(10):1893–1915
10. Rossman LA (2010) Storm water management model user's manual. Version

11. Sahoo SN, Sreeja P (2014) A methodology for determining runoff based on imperviousness in an ungauged peri-urban catchment. Urban Water J 11(1):42–54
12. Sahoo SN, Sreeja P (2016) Determination of effective impervious area for an urban Indian catchment. J Hydrol Eng Am Soc Civil Eng 21(4):5016004
13. Tabari H, De Troch R, Giot O, Hamdi R, Termonia P, Saeed S, Brisson E, Van Lipzig N, Willems P (2016) Local impact analysis of climate change on precipitation extremes: are high-resolution climate models needed for realistic simulations? Hydrol Earth Syst Sci 20(9):3843–3857
14. Willems P, Vrac M (2011) Statistical precipitation downscaling for small-scale hydrological impact investigations of climate change. J Hydrol 402(3–4):193–205
15. Zahmatkesh Z, Burian SJ, Karamouz M, Tavakol-Davani H, Goharian E (2015) Low-impact development practices to mitigate climate change effects on urban stormwater runoff: case study of New York City. J Irrig Drain Eng 141(1):04014043

Check for updates

Chapter 2
Impact of Climate Change on Daily Maximum Temperature of the Brahmaputra River Basin

Pulendra Dutta[1](✉) and Arup Kumar Sarma[2]

[1] Department of Civil Engineering, Jorhat Engineering College, Jorhat 785007, India
[2] Department of Civil Engineering, Indian Institute of Technology, Guwahati 781039, India

Abstract. The climate change across the globe basically happens due to the emissions of the greenhouse gases. Climate change is linked to the changes in both the temperature and precipitation cycles. These changes lead to the spatial and temporal variations of the global precipitation patterns. Consequently, the water resources are affected in terms of both the quantity and quality of natural water bodies. The Brahmaputra River Basin having wide spatial variations in topography as well as weather components is also susceptible to the impacts of global climate change. The present study is forwarded to assess the future climate change of this mighty river basin with special emphasis on the daily maximum temperature (TMax) during 2020–2100. Initially, the interpolated and bias corrected historical variables of several GCMs were analyzed to identify one GCM suitable for the present area. Finally, Mann–Kendall test and Sen's slope method were applied to determine the trend in the future TMax. The results show that the mean value of the daily temperature would increase by 0.08 °C per year.

Keywords: Brahmaputra river basin · Climate change · Maximum temperature

1 Introduction

Climate change due to the increase of greenhouse gas emissions is considered to be one of the major challenges to human beings in the twenty-first century. As per the fourth assessment report (AR4) of Intergovernmental Panel on Climate Change (IPCC), the earth's surface temperature would increase by 1.1–6.4 °C toward the end of the twenty-first century. Moreover, there would be a significant rise in global mean sea level. These changes would lead to changes in the hydrological cycle which ultimately results in changes in atmospheric moisture, precipitation, evaporation, etc. Even it would catch raising frequencies of the extreme events of a catchment area. Thus, the human society will be influenced in manyfold such as decrease in agricultural production, increase in the risks of lives, socioeconomic harm, rise in water conflicts, and poverty. However, the risk and cost by and large would be equivalent to significant loss of the global gross

Pulendra Dutta—Presently pursuing PhD at IIT Guwahati

R. K. Bhattacharjya et al. (eds.), *Sustainable Water Resources Management*, Advances in Sustainability Science and Technology,
https://doi.org/10.1007/978-981-16-7535-5_2

domestic product (GDP). As such, it is important to accurately estimate the amount of variability due to climate change of the natural factors especially in the hydrological cycle and flooding events. This will present a strong root for justifying the impacts of climate change and to adopt suitable strategies.

The variability of hydrological variables can produce natural disasters such as flooding or drought and can threaten our natural and social ecosystems. Hydrological cycle components such as precipitation, temperature, and evapotranspiration are impacted by the global climate change. There have been several studies [4, 13] worldwide for detecting and reporting the changes in the hydrological variables. Chen et al. [2] reported how streamflow of Yangtze River changes due to the changes in precipitation because of climate change. Similarly, other variables need to be addressed for the impact on basin behavior. Especially, temperature being very sensitive for a glacier-fed area like the Brahmaputra River Basin should be addressed for its variability that would ultimately result in hydraulic alterations. In a study, Barman and Bhattacharjya [1] utilized global climate model data to articulate the changes in snow cover over the Brahmaputra River Basin due to rise in temperature, whereas the information provided by such researches can be helpful for water resource management of a river basin. It is however difficult to build a consensus about how the changes in climatic variables are taking place. This is because of lot of uncertainties involved in the climate change studies. It basically includes the climate models that have a lot of uncertainties due to limited surface data and complex watershed properties. The Brahmaputra River Basin occupying a huge area (nearly 5.8 lacs km^2) has a very unique and complex characteristic. Due to its complex characteristics and high weather as well as topographical variability, the Brahmaputra River Basin poses a huge challenge to the research community. This mighty river basin has hardly been addressed as a single entity considering the entire river stretch starting from its source at Tibet to the end at Bangladesh. Although several case-specific studies [3, 9, 10] have been forwarded on a particular area of the basin, none of the above studies forwarded changes in climatic variable and its subsequent impact analysis on the basin under consideration.

The mechanism by which the changes of hydrological variables are being driven by changes in climate variables should be analyzed. The study of the impact of climate change on hydrological variables depends on projections of potential climate provided by various general circulation models (GCMs). But we hardly use the direct outputs of GCM because they produce error/biases. This error may rise due to the wider spatial resolution and diverse thermodynamic and climate scheme processes [6]. Otherwise, their direct use would lead to error with respect to historical observations [8]. As such, downscaling and bias corrections of the GCM data become inevitable prior to its application for a certain area. Downscaling of GCM models is a challenging task especially for the large Brahmaputra River Basin having limited information. The main objective of this study is to project the future variation of climatic variables, especially the daily maximum temperature. For this, the GCM outputs are interpolated and then bias corrected in order to predict the likely maximum temperature of the Brahmaputra River Basin up to 2100.

2 Study Area

The transboundary Brahmaputra River Basin (Fig. 1) has been taken up for the present study. This basin includes four nations such as China, India, Bhutan, and Bangladesh. Originating in Southern Tibet at an elevation of 5300 m, the Brahmaputra River ends its journey at the Bay of Bengal in Bangladesh. The Brahmaputra flows for about 2880 km, and it covers a major journey in Tibet. Then, it flows through India to finally end its journey at the Bay of Bengal in Bangladesh. The major part of the basin area is occupied by China (~50%), whereas India occupies about 36%, and Bhutan and Bangladesh share nearly equal (~7%). The basin is characterized by a wide spatial variability of weather components, land use, and topography. The temperature across the basin has a large variability, and its value ranges from very low at the Himalayan range to 35–39 °C during summer in the plain areas. Spreading over more than 580,000 km^2, this river basin is also characterized by a wide spatial variation of flora and fauna.

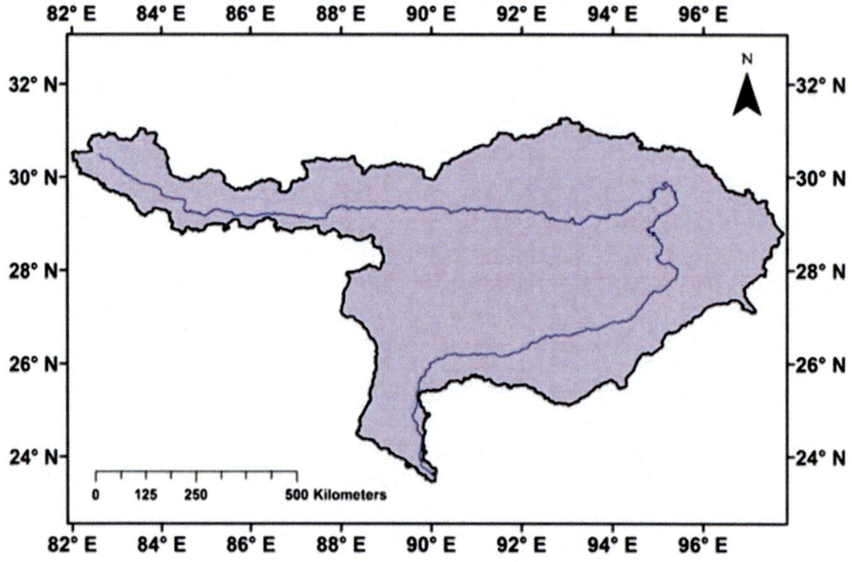

Fig. 1 Study area, i.e., the Brahmaputra River basin

3 Data

The Shuttle Radar Transmission Mission (SRTM) digital elevation model (DEM) of 90 m spatial resolution is used for delineating the river basin. As for the observed weather data are concerned, we have utilized several sets of data collected from various sources locally and globally. The local data include station data at Indian Meteorological Department (IMD) and tea estates of Assam, while the global data include the IMD gridded data and the Climate Forecast System Reanalysis (CFSR) weather data provided by Texas

A&M University (TAMU). Three CMIP5 GCMs, viz. GFDL-ESM2M, HadGEM2-CC, and IPSL-CM5-1R, were used in the present study. These GCMs were selected because the climate data for the study periods spanning from the present till future are available. These GCM data are freely available at Web sources.

4 Methods

Initially, DEM is used to delineate the boundary of the basin. The Brahmaputra River Basin is subdivided into 36 sub-basins to account for all the major tributaries. Out of numerous available observed weather records, we selected only 36 stations' data so that each sub-basin has at least one set of station data. Simulated daily maximum temperature (TMax) values were obtained from the three GCMs for the period 1991–2005, each with different grid sizes. Knowing the coordinates of 36 stations and that of different grids, the GCMs output was interpolated in order to obtain the TMax at each weather station location. Here, we used the inverse distance weighted average (IDWA) interpolation method, the most widely adopted process worldwide, which assumes that the effect of neighboring points on the desired interpolated point solely depends on the inverse of the distance between the neighboring points and the desired location.

Bias correction was carried out by employing the linear scaling method [12] on the interpolated daily TMax. The bias correction factors obtained from the analysis are then applied to the climate data of the GCM, and bias-corrected climate variables for the present (1991–2005) periods are thus obtained. Applying certain statistical parameters, the performance of the GCMs was evaluated to identify one GCM best suitable for the present area. Here, we found that HadGEM2-CC outshines the other two, and as such, we forwarded this GCM only for the future (2020–2100) periods. Here, the bias correction factors obtained for the historical period (1991–2005) were kept same and hence applied for correcting the future (2020–2100) variable of the GCM. Finally, the trend analysis for the TMax variable was carried out by using nonparametric Mann–Kendal test trend [5, 7] analysis. Here, the significance of the trend values was tested by Sen's slope method [11]. The trend analysis was carried out for the annual mean values of the daily TMax.

5 Results and Discussion

Table 1 shows the results of GCM comparison using certain statistical measures widely adopted by the researchers across the globe. The parameters such as root mean square error (RMSE) and Nash–Sutcliffe (NS) coefficients are shown for all the GCMs [i.e., GFDL-ESM2M, HadGEM-2CC, IPSL-CM5-LR] considered in this study. It is evident from this table that the values of RMSE and NS of the HadGEM-2CC model are minimum as compared to the other GCMs. For example, the mean RMSE value (=3.33) of this GCM is minimum than the corresponding values of GFDL-ESM2M (=3.75) and IPSL-CM5-LR (=3.54). This means that the bias-corrected GCM data show more resemblance with the observed weather data, for HadGEM-2CC, than the other GCMs. As such, HadGEM-2CC is considered to be the best GCM for the present area and forwarded for the future analyses.

Table 1 RMSE and NSE values of maximum temperature (°C) of different GCMs

Index	Component	GCM1	GCM2	GCM3
RMSE	Mean	3.75	**3.33**	3.54
	Standard deviation	0.47	**0.39**	0.39
	Median	3.81	3.41	3.63
NSE	Mean	0.48	**0.46**	0.55
	Standard deviation	0.25	**0.16**	0.22
	Median	0.49	0.58	0.56

Note GCM-1: GFDL-ESM2M; GCM2: HadGEM2-CC; GCM3: IPSL-CM5-LR
The bold display refers to the best-fitted values

The variation of annual mean maximum temperature is shown in Fig. 2. Here, the TMax value of HadGEM2-CC at a station is obtained by averaging the daily maximum temperature values of a year and over the entire periods. The entire future periods (2020–2100) is divided into three sub-periods as F1 (2020–2040), F2 (2041–2070), and F3 (2071–2100). The mean annual maximum temperature during 2020–2040 (Fig. 2a) seemed to widely vary over the study area. It is observed that Sen's slope value stands the highest (9.8) at the middle reach of the China part of the Brahmaputra River Basin. This indicates the projected maximum temperature at this location would increase by 9.8%. Overall, the TMax of the entire basin seems to increase up to 2040. Moreover, these trend values are found significant (S) at majority of the stations, as tested by the Mann–Kendall test. On the other hand, these values at a few stations are found non-significant (NS) which may be due to inability of the GCM while estimating the weather variables for an area having very complex topographic characteristics. Moreover, bias may also be another concern. It is to note that none of the bias correcting methods is capable to completely remove the biases. So, it may be concluded that the climate projection represents a fair estimation for the present study. Similar phenomena have also been observed for the other future periods (F2 and F3). The increase in annual mean TMax would happen up to 12.8% (Fig. 2b) and 10.6% (Fig. 2c) during 2070 and 2100, respectively. Thus, the maximum temperature of the Brahmaputra River Basin would increase by 0.08 °C/year (Fig. 3) up to 2100. As a result, the Brahmaputra River Basin would experience climate change in the coming days which would ultimately impact the basin behavior in terms of hydraulics, hydrology, and eco-system.

6 Conclusions

In climate change studies, downscaling of GCM outputs plays an important role since the grid points of the GCM hardly coincide with the observed weather station. However, we applied only the interpolation of these data followed by the bias correction. This is because the selected GCM stations remained at close proximity to the observed station. The future mean maximum basin temperature is likely to increase by 0.08 °C/year.

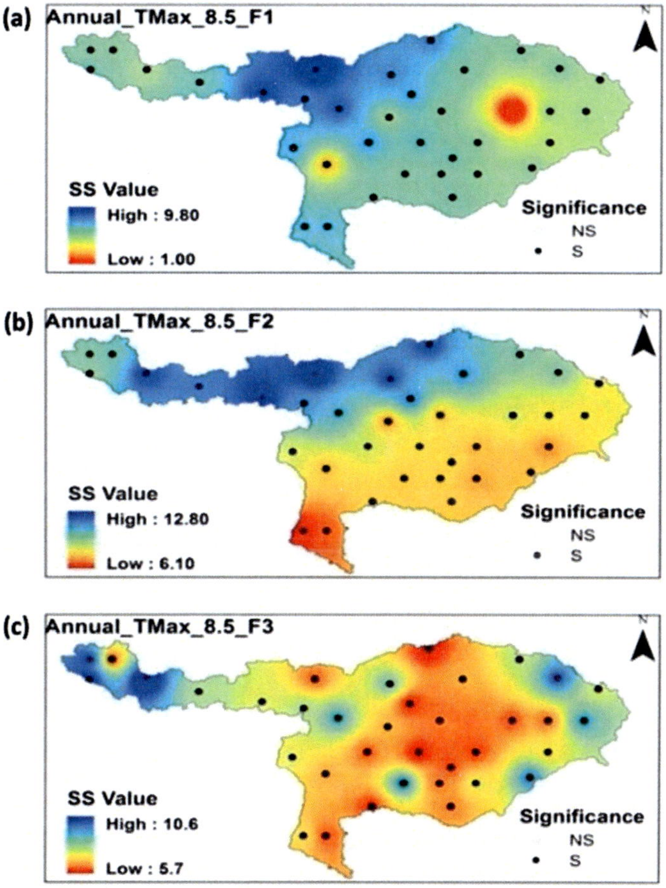

Fig. 2 Variation of annual mean maximum temperature for the Brahmaputra River Basin during: **a** Future 1 (F1) [2020–2040]; **b** Future 2 (F2) [2041–2070]; and **c** Future 3 (F3) [2071–2100]

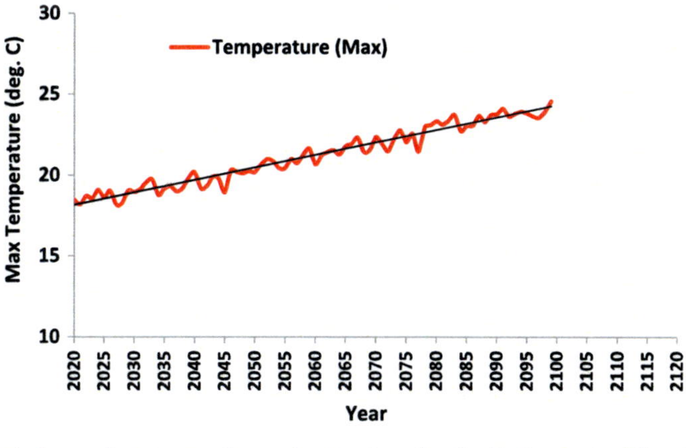

Fig. 3 Projected annual mean maximum temperature for the Brahmaputra River Basin during 2020–2100

However, this value may vary little if we could have considered more than three GCMs considered in the present study. And, this may be forwarded as a scope of the future study.

References

1. Barman S, Bhattacharjya RK (2015) Change in snow cover area of Brahmaputra river basin and its sensitivity to temperature. J Environ Syst Res. https://doi.org/10.1186/s40068-015-0043-0
2. Chen J, Wu X, Finlayson BL et al (2014) Variability and trend in the hydrology of the Yangtze River, China: annual precipitation and runoff. J Hydrol 513:403–412
3. Goswami DC (1985) Brahmaputra River, Assam, India: physiography, basin denudation, and channel aggradation. Water Resour Res 21:959–978
4. Ji X, Lesack LFW, Melack JM et al (2019) Seasonal and interannual patterns and controls of hydrological fluxes in an amazon floodplain lake with a surface-subsurface process model. Water Resour Res
5. Kendall MG (1975) Rank correlation methods, 4th edn. Charles Griffin, London
6. Lazoglou G, Anagnostopoulou C, Skoulikaris C, Tolika K (2019) (2019) Bias correction of climate model's precipitation using the Copula method and its application in river basin simulation. Water 11:600
7. Mann HB (1945) Nonparametric tests against trend. Econometrica 13:245–259
8. Ramirez-Villegas J, Challinor AJ, Thornton PK, Jarvis A (2013) Implications of regional improvement in global climate models for agricultural impact research. Environ Res Lett 8:24018
9. Sahoo SN, Sreeja P (2015) Development of flood inundation maps and quantification of flood risk in an urban catchment of Brahmaputra River. ASCE-ASME J Risk Uncertain Eng Syst Part A Civ Eng 3:A4015001
10. Sarma JN (2005) Fluvial process and morphology of the Brahmaputra River in Assam, India. Geomorphology 70:226–256
11. Sen PK (1968) Estimates of the regression coefficient based on Kendall's Tau. J Am Stat Assoc 63:1379–1389. https://doi.org/10.1080/01621459.1968.10480934
12. Shrestha M, Acharya SC, Shrestha PK (2017) Bias correction of climate models for hydrological modelling–are simple methods still useful? Meteorol Appl 24:531–539
13. Skoulikaris C, Ganoulis J (2012) Climate change impacts on river catchment hydrology using dynamic downscaling of global climate models. In: Fernando H et al (eds) National security and human health implications of climate change, NATO Science for Peace and Security Series C: Environmental Security, pp 281–287. https://doi.org/10.1007/978-94-007-2430-3_240

Chapter 3
Future Rainfall Trend Analysis Over Puthimari River Basin: A Comparative Study Using Different CMIP5 Models

Swapnali Barman[1](✉), Jaivir Tyagi[2], and Waikhom Rahul Singh[1]

[1] Center for Flood Management Studies, Guwahati, National Institute of Hydrology, Dispur 781006, India
wrsingh.nihr@gov.in
[2] National Institute of Hydrology, Roorkee, Roorkee 247667, India
tyagi.nihr@gov.in

Abstract. River Puthimari is a north bank tributary of River Brahmaputra. The basin, being a part of the Assam valley, is subjected to heavy rainfall especially during the monsoon season, i.e., from June to September. To analyze the response of rainfall to climate change over Puthimari basin, the present study considers five different CMIP5 models for RCP4.5 and RCP8.5, and the future rainfall up to 2099 has been analyzed using Mann–Kendall non-parametric test. Nature of rainfall trends has been identified by examining the Z statistics for each of the model. The results obtained from different models were different indicating that a single model analysis is not enough to understand the future rainfall pattern over the basin. However, from the general interpretation of all the models, it has been observed that rainfall under RCP8.5 will be more compared to that under RCP4.5 and the historical period. There is also a possibility of decrease of rainfall in the monsoon season toward the end of the century.

Keywords: Rainfall · Climate change · CMIP5 · Mann–Kendall · RCP

1 Introduction

Accelerated change in climate is expected to have major impact on the future sustainability of the earth [1]. Increase in anthropogenic emissions of gases (e.g., carbon dioxide, methane, etc.) into the atmosphere and an enhanced greenhouse effect are the major driving forces behind the accelerated global warming that has taken place over the last century [2]. Climate change is assessed, and climate projection is produced efficiently using the global climate models (GCMs) [14'. With higher spatial resolution and smaller bias compared to Coupled Model Intercomparison Project Phase 3 (CMIP3) models, in the models provided by the CMIP5, carbon cycle and dynamic vegetation modules have been incorporated, and the mechanisms involved are quite complicated that provides better simulation ability of some climate features [3–7]. The climate simulations of

R. K. Bhattacharjya et al. (eds.), *Sustainable Water Resources Management*,
Advances in Sustainability Science and Technology,
https://doi.org/10.1007/978-981-16-7535-5_3

CMIP5 models in AR5 have been done according to representative concentration pathways (RCPs) [8]. RCPs are based on four greenhouse gas concentration trajectories, i.e., 2.6, 4.5, 6 and 8.5 W/m^2 that indicate radiative forcing for each RCP [9, 10]. One of the major impacts of climate change is variation in rainfall pattern which directly or indirectly affects the regional water sources [11]. Emission of the anthropogenic greenhouse gases and associated raise in temperature is responsible for increase in the intensity of precipitation [2]. According to Shiu et al. (2012), the accumulated atmospheric moisture in high amount due to warming climate is the reason behind increased precipitation [12]. Again, a study by Dore (2005) revealed that the wet areas have become wetter, and arid areas have become drier due to climate change [13]. Mohapatra et al. (2018) from their study of linear trend analysis of Indian Summer Monsoon Rainfall (ISMR) observed decreasing trend of rainfall in Northeast India and increasing trend in Northwest India over the last two decades [14]. However, there is a possibility that under global warming and climate change, the precipitation may increase at high latitudes and in the vicinity of the equator but decrease in the subtropics [15]. Again, in East Asia, South Asia and Southeast Asia, precipitation may increase in future due to climate change which in turn will increase flood prone areas in these regions [16]. In case of the Indian subcontinent, the impact of climate change will mainly on the hydrology in this region as climate change will cause shifting of rain pattern across India because of which many areas will suffer from flood and many others will suffer from draft [17]. It is expected that the hydrological behavior of the Ganges–Brahmaputra basin under climate change will be stronger compared to the other major river systems of the world. Particularly in the Brahmaputra basin, which is mainly influenced by the monsoon rainfall, climate change will affect both high and low flows leading to variability of available water both in space and time [18]. The Brahmaputra River receives many tributaries during its journey from the Chema Yung Dung Glacier in Southern Tibet up to the Bay of Bengal. Among these tributaries, Puthimari River which originates in the Bhutan Himalaya [19] is considered as one of the major north bank tributaries of Brahmaputra. Puthimari basin lies between 26° 10′ N–27° 18′ N latitude and 91° 27′ E–91° 50′ E longitude covering a total area of 1500 km^2. Almost half of the basin falls in the State of Assam in India. Assam, especially Kamrup, and its nearby districts experience heavy annual average rainfall of approximately 1700 mm [19]. Such heavy rainfall in the monsoon season results in heavy discharge in the river which in turn leads to problems like floods and erosion in the downstream causing huge losses [1]. In the present study, change in rainfall over Puthimari basin due to climate change has been analyzed up to the year 2099. Mann–Kendall test has been performed to examine the rainfall trends for five different CMIP5 models under RCP4.5 and RCP8.5 (Fig. 1).

2 Materials and Methodology

Data Used

Bias corrected downscaled rainfall data available at 25 km × 25 km spatial resolution from five different CMIP5 models for RCP4.5 and RCP8.5 have been used in this study. Table 1 gives the names of the models and their respective resolutions.

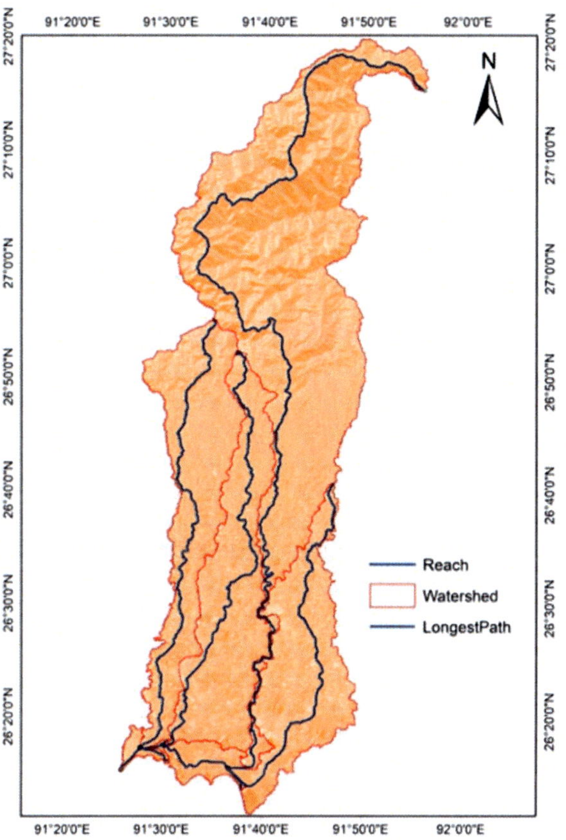

Fig. 1 Puthimari basin

Table1 CMIP5 models and their resolutions

Sl. No	Model	Latitude	Longitude
1	ACCESS1.0	1.25	1.875
2	BCC-CSM1.1	2.7906	2.8125
3	CCSM4	0.9424	1.25
4	CanESM2	2.7906	2.8125
5	IPSL-CM5A-MR	1.2676	2.5

Methodology

The impact of climate change on rainfall over Puthimari basin has been analyzed up to 2099 and has been compared to the historical rainfall. Where the historical rainfall has been taken from 1970 to 2005, the future rainfall has been analyzed for three different time periods viz., 2025–49, 2050–74 and 2075–99. Mann–Kendall [3, 4] is a non-parametric test used to identify trends in time series data. In this study, Mann–Kendall test has been performed for each CMIP5 model and for each time period to determine the trends in monthly rainfall. The statistical significance of the trends has been identified by analyzing the Z values. A positive value of Z indicates an upward trend while a negative value of Z indicates a downward trend.

The Mann–Kendall test is calculated as,

$$S = \sum_{i=1}^{N-1} \sum_{j=i+1}^{N} \operatorname{sgn}(x_j - x_i) \tag{1}$$

where N is the number of data points, and x_i and x_j are the data values in the time series i and j, respectively. sgnθ is determined as,

$$\operatorname{sgn}(x_j - x_i) = \begin{cases} 1 & \text{if} x_j - x_i > 0 \\ 0 & \text{if} x_j - x_i = 1 \\ -1 & \text{if} x_j - x_i < 1 \end{cases} \tag{2}$$

The variance is calculated as,

$$\operatorname{Var}(S) = \frac{N(N-1)(2N+5) - \sum_{k=1}^{n} t_k(t_k - 1)(2t_k + 5)}{18} \tag{3}$$

where n is the number of tied groups, and t_k is the number of data points in the kth tied group.

The standard Z statistics is then calculated as,

$$Z = \begin{cases} \frac{S-1}{\sqrt{\operatorname{Var}(S)}} & \text{if } S > 0 \\ 0 & \text{if } S = 0 \\ \frac{S+1}{\sqrt{\operatorname{Var}(S)}} & \text{if } S < 0 \end{cases} \tag{4}$$

If Z value lies between ±1.96, the null hypothesis of having no trend in the series cannot be rejected at 95% level of confidence.

3 Results and Discussions

The monthly rainfall variations under RCP4.5 and RCP8.5 for each of the model are plotted in Figs. 2 and 3, respectively.

From the figures, it is clear that rainfall will not increase continuously from 2025 to 2099. Different models show different variations under the two RCPs. For example, under both the RCPs, for most of the months, ACCESS model shows increase in rainfall in future compared to the historical period. However, during 2025–49, rainfall in June, September and October is less than the historical period under RCP4.5. Under RCP8.5, maximum average rainfall of 3339.88 mm is observed in the month of June during the period 2025–49. For BCC-CSM1.1, maximum rainfall of 316.96 and 418.18 mm is observed in July under RCP4.5 (1970–2005) and RCP8.5 (2075–99), respectively. More

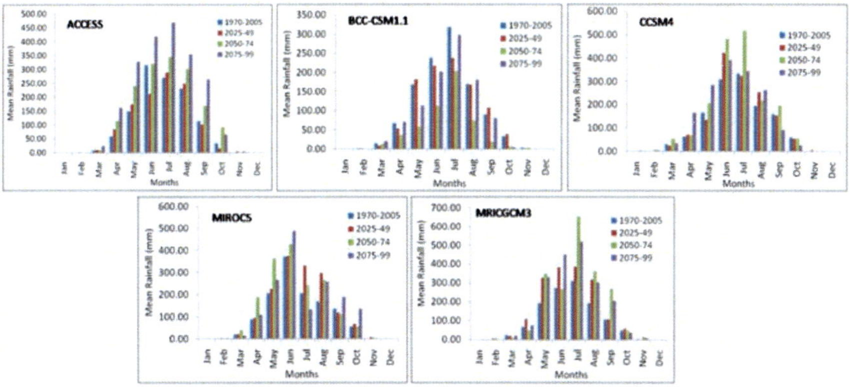

Fig. 2 Monthly rainfall variations of different models under RCP 4.5

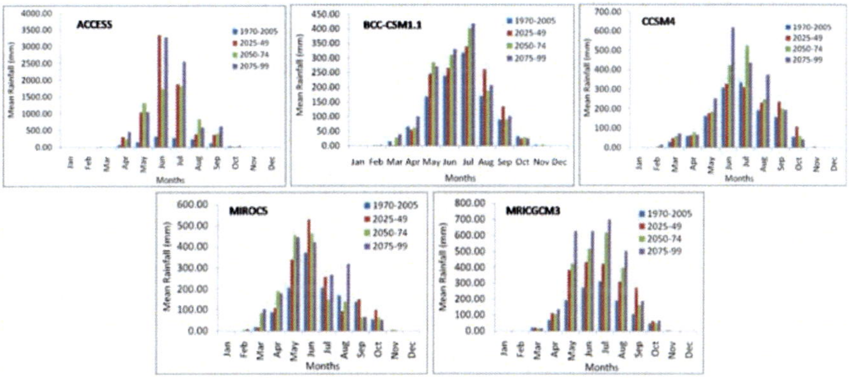

Fig. 3 Monthly rainfall variations of different models under RCP 8.5

rainfall in August and September during 2025–49 has been observed compared to the later time periods. Toward the end period, i.e., during 2075–99, rainfall in September and October is the minimum for CCSM4 model under RCP4.5, and the maximum rainfall of 517.02 mm is observed in July during 2050–74. Under RCP8.5, maximum rainfall of 617.93 mm is observed in June during 2075–99. For MIROC model, in the month of July, the rainfall will decrease to 132.71 mm in 2075–99 under RCP4.5. However, higher average rainfall has been observed in April and May during 2050–74, and in July and August during 2025–49 compared to rainfall during 2075–99. Rainfall in July during 2050–74, August during 2025–49 and 2050–74, September during 2075–99 and October during 2075–99 are less compared to the historical period under RCP8.5. The maximum rainfall of 487.30 mm (2075–99) and 528.08 mm (2025–49) is observed in the month of June under RCP4.5 and RCP8.5, respectively. For the fifth model, i.e., MRICGCM3, under RCP4.5, rainfall in July, August and September during the period 2050–74 is found to be more compared to other time periods. Maximum rainfall of 650.88 mm is observed in July. Except in the month of September, where rainfall during 2025–49 is more than other time periods, for rest of the months, rainfall increases continuously toward the end of the century, and the maximum rainfall of 698.11 mm is observed in July during 2075–99.

The nature of the trends of rainfall variation for each month in different time periods under the two RCPs is determined by analyzing the Z statistics obtained from the Mann–Kendall test. The heat maps prepared for the Z values for each of the five models are shown in Figs. 4 and 5 for RCP4.5 and RCP8.5, respectively. The nature of trends obtained for a particular month is not the same for each model. Under RCP4.5, during 2025–49, majority of the models show decreasing trends in the months from May to August. Maximum positive Z value is obtained in May for CanESM, and maximum negative Z value is obtained in February for MIROC. The same MIROC model shows maximum positive trend in February during 2050–74, and maximum negative trend is observed for BCC-CSM1 in the month of July. Most of the models show either negative trends or no significant trend in the months from May to August. Except ACCESS (July), CCSM4 (April, May, June, August), MIROC (July) and MRICGCM3 (June, July), the other show either negative or no significant trend during 2050–74 in the months from May to August. During the period 2075–99, maximum positive and negative values are observed in July and April, respectively, for MRICGCM3.

Month	ACCESS				CCSM4				MIROC				MRICGCM3				BCCCSM1			
	1970-2005	2025-49	2050-74	2075-99	1970-2005	2025-49	2050-74	2075-99	1970-2005	2025-49	2050-74	2075-99	1970-2005	2025-49	2050-74	2075-99	1970-2005	2025-49	2050-74	2075-99
Jan	0	0	0	0	0	0	0	-1.17	0	0	0	0	0	0	0	0	0	0	0	0
Feb	1.76	-0.78	0	1.56	0	1.56	0	2.13	-0.46	-2.56	3	1.05	0	-1.89	-0.36	0	0.23	1.37	0	0
Mar	-0.33	-0.96	-1.78	-0.88	-0.09	0.38	-1.46	-0.62	1.83	-2.27	-2.76	-0.11	-0.7	0.34	1.01	2.39	0.86	-0.05	-2.44	0.7
Apr	-1.02	2.81	-1.2	-1.76	0.33	-1.9	0.67	0.81	2.13	1	0.23	1.38	-1.96	-0.7	-1.14	-2.09	1.7	-0.16	-4.37	1.3
May	-0.55	-1.25	0.11	1.94	0.46	0.12	2.01	2.16	0.51	1.31	-0.06	0.93	0.67	-1.03	-4.14	-1.31	-0.12	-0.38	-4.15	-0.1
Jun	0.07	-1.19	-1.48	-1.58	-0.7	1.2	2.47	-0.36	-0.6	0	-0.87	0.69	1.02	-1.23	0.92	-1.73	-0.03	-1.16	-4.61	-0.01
Jul	-0.97	-0.25	1.35	-0.61	-0.39	1.33	-3.05	-1.07	2.84	0.08	1.77	-0.8	-0.83	-0.31	0.44	3.99	1.83	-0.03	-5.02	1.7
Aug	-0.27	-0.92	0.31	1.67	-0.72	-1.01	1.35	0.17	0.78	2.68	-0.65	1.59	-2.02	-1.82	-1.71	0	0.17	-0.65	-4.39	-0.5
Sep	-0.19	2.51	-1.33	-1.84	-2.21	0.51	0.3	-0.71	-1.46	1.86	-0.62	0.94	0.11	1.68	0	-0.79	-0.22	1.2	-3.19	0.2
Oct	-0.35	-1.64	-0.6	1.75	-1.27	-0.47	0	0.63	0.71	-1.53	0.62	-0.2	0.9	-0.73	-0.03	0.83	-1.54	-0.21	-1.96	-1.1
Nov	-0.47	1.05	0	1.56	0.64	-0.98	0	0	-0.09	-2.03	2.31	-1.17	1.17	0	0.14	1.62	0.96	0	0.78	0.9
Dec	0	0	0	0	0	0	0	0	0.28	0	0	0	0	0	0	0	0	0	0	0

Min Max

Fig. 4 Heat maps showing Z values under RCP4.5

Under RCP8.5, during 2025–49, maximum positive and negative trends are observed in August and September for ACCESS and BCC-CSM1, respectively. Maximum positive and negative trends in July and September, respectively, are observed during 2050–74 for the ACCESS model. During 2075–99, the maximum positive and negative trends are observed in April and June, respectively, for BCC-CSM1.

The positive and negative trends of average monthly rainfall for each model for different time periods are symbolically represented in Figs. 6 and 7 for RCP4.5 and RCP8.5, respectively.

Under RCP4.5, for ACCESS, trends remain the same for January(No trend), March(-ve), June(-ve), November(+ve) and December(no trend) from 2025 to 99. Positive trends are seen toward the end period for February, May, August and November. For April and September, positive trends during 2025–49 become negative from 2050 to 99. For July, alternate trends of negative, positive and negative are observed for the three time periods.

Month	ACCESS				CCSM4				MIROC				MRICGCM3				BCCCSM1			
	1970-2005	2025-49	2050-74	2075-99	1970-2005	2025-49	2050-74	2075-99	1970-2005	2025-49	2050-74	2075-99	1970-2005	2025-49	2050-74	2075-99	1970-2005	2025-49	2050-74	2075-99
Jan	0	0	0	0	0	0	0	1.98	0	0	0	1.76	0	0	0	0	0	0	0	0
Feb	1.76	0	0	0	0	1.37	0.41	-0.41	-0.46	-0.78	0.39	0.71	0	0	-1.41	-0.85	0.23	2.35	1.37	-0.78
Mar	-0.33	0	-0.85	-0.78	-0.09	0.72	0.91	-1.76	1.83	-0.17	-0.83	0.55	-0.7	-0.6	-0.84	-0.52	0.86	0.28	0.56	-1.87
Apr	-1.02	-0.48	-0.31	-0.45	0.33	1.84	-1.7	-1.07	2.13	0.83	2.06	1.46	-1.96	-0.18	2	2.85	1.7	3.42	-0.16	3.1
May	-0.55	0.64	-0.22	1.86	0.46	0.9	-0.11	0.65	0.51	2.58	0.86	-0.47	0.67	-1.48	1.76	-0.17	-0.12	-0.56	-0.4	1.71
Jun	0.07	1.22	1.66	0.42	-0.7	0.48	-0.39	-0.28	-0.6	0.36	0.06	0.34	1.02	0.03	0.78	-1.73	-0.03	0.78	-1.68	-2.23
Jul	-0.97	-0.06	4.07	1.25	-0.39	-1.89	1.36	0.81	2.84	0.74	-1.09	1.01	-0.83	0.56	-0.89	-0.61	1.83	1.76	-0.56	1.99
Aug	-0.27	3.47	-0.28	1.55	-0.72	-0.62	2.39	-0.69	0.78	0.26	0.17	-0.23	-2.02	0.03	1.45	-1.19	0.17	-0.08	1.33	-0.44
Sep	-0.19	1.2	0.45	1.78	-2.21	0.7	-2.24	1.74	-1.46	-0.95	-0.8	-0.62	0.11	0.14	2.61	-0.6	-0.22	-3.22	-0.33	0.75
Oct	-0.35	-1.72	-2.48	1.52	-1.27	-0.33	0.36	1.91	0.71	0.24	-0.59	0.08	0.9	0.36	-0.22	-1.67	-1.54	0.04	1.74	1.27
Nov	-0.47	0	0	0	0.64	-2.44	1.17	0	-0.09	0.07	1.47	1.17	1.17	1.75	0.78	0.07	0.96	0	0.59	0
Dec	0	0	0	0	0	0	0	0	0.28	0	0	0	0	0	0	0	0	0	0	0

Min Max

Fig. 5 Heat maps showing Z values under RCP8.5

Month	ACCESS				CCSM4				MIROC				MRICGCM3				BCCCSM1			
	1970-05	2025-49	2050-74	2075-99	1970-05	2025-49	2050-74	2075-99	1970-05	2025-49	2050-74	2075-99	1970-05	2025-49	2050-74	2075-99	1970-05	2025-49	2050-74	2075-99
Jan	→	→	→	→	→	→	→	↓	→	→	→	→	→	→	→	→	→	→	→	→
Feb	↑	↓	→	↑	→	↑	→	↑	↓	↓	↑	↑	→	↓	↓	→	↑	↑	→	→
Mar	↓	↓	↓	↓	↓	↑	↓	↓	↑	↓	↓	↓	↓	↑	↑	↑	↑	↓	↓	↑
Apr	↓	↑	↓	↓	↑	↓	↑	↑	↑	↑	↑	↑	↓	↓	↓	↓	↑	↓	↓	↑
May	↓	↓	↑	↑	↑	↑	↑	↑	↑	↑	↓	↑	↑	↓	↓	↓	↓	↓	↓	↓
Jun	↑	↓	↓	↓	↓	↑	↑	↓	↓	→	↓	↑	↑	↓	↑	↓	↓	↓	↓	↓
Jul	↓	↓	↑	↓	↓	↑	↓	↓	↑	↑	↑	↓	↓	↓	↑	↑	↑	↓	↓	↑
Aug	↓	↓	↑	↑	↓	↓	↑	↑	↑	↑	↓	↑	↓	↓	↓	→	↑	↓	↓	↓
Sep	↓	↑	↓	↓	↓	↑	↑	↓	↓	↑	↓	↑	↑	↑	→	↓	↓	↑	↓	↑
Oct	↓	↓	↓	↑	↓	↓	→	↑	↑	↓	↑	↓	↑	↓	↓	↑	↓	↓	↓	↓
Nov	↓	↑	→	↑	↑	↓	→	→	↓	↓	↑	↓	↑	→	↑	↑	↑	→	↓	↑
Dec	→	→	→	→	→	→	→	→	↑	→	→	→	→	→	→	→	→	→	→	→

Fig. 6 Nature of trends of monthly rainfall of the CMIP5 models for different time periods under RCP4.5

Month	ACCESS				CCSM4				MIROC				MRICGCM3				BCCCSM1			
	1970-05	2025-49	2050-74	2075-99	1970-05	2025-49	2050-74	2075-99	1970-05	2025-49	2050-74	2075-99	1970-05	2025-49	2050-74	2075-99	1970-05	2025-49	2050-74	2075-99
Jan	→	→	→	→	→	→	→	↑	→	→	→	↑	→	→	→	→	→	→	→	→
Feb	↑	→	→	→	→	↑	↑	↓	↓	↓	↑	↑	→	→	↓	↓	↑	↑	↑	↓
Mar	↓	→	↓	↓	↓	↑	↑	↓	↑	↓	↓	↑	↓	↓	↓	↓	↑	↑	↑	↓
Apr	↓	↓	↓	↓	↑	↑	↓	↓	↑	↑	↑	↑	↓	↓	↑	↑	↑	↑	↓	↑
May	↓	↑	↓	↑	↑	↑	↓	↑	↑	↑	↑	↓	↑	↓	↑	↓	↓	↓	↓	↑
Jun	↑	↑	↑	↑	↓	↑	↓	↓	↓	↑	↑	↑	↑	↑	↑	↓	↓	↑	↓	↓
Jul	↓	↓	↑	↑	↓	↓	↑	↑	↑	↑	↓	↑	↓	↑	↓	↓	↑	↑	↓	↑
Aug	↓	↑	↓	↑	↓	↓	↑	↓	↑	↑	↑	↓	↓	↑	↑	↓	↑	↓	↑	↓
Sep	↓	↑	↑	↑	↓	↑	↓	↑	↓	↓	↓	↓	↑	↑	↑	↓	↓	↓	↓	↑
Oct	↓	↓	↓	↑	↓	↓	↑	↑	↑	↑	↓	↑	↑	↑	↓	↓	↓	↑	↑	↑
Nov	↓	→	→	→	↑	↓	↑	→	↓	↑	↑	↑	↑	↑	↑	↑	↑	→	↑	→
Dec	→	→	→	→	→	→	→	→	↑	→	→	→	→	→	→	→	→	→	→	→

Fig. 7 Nature of trends of monthly rainfall of the CMIP5 models for different time periods under RCP8.5

For the month of October, for the first two time periods, rainfall shows negative trends, and toward the end of the century, it becomes positive. For CCSM4, there is a possibility that rainfall will decrease for January, March and July from 2050 to 99. Again, positive trends are seen from 2050–99 for February, April, May, August, October and November. However, for June and September, after continuous positive trends from 2025 to 74, it becomes negative from 2075 to 99. In case of MIROC model, for May, June, August and September, positive trends during 2025–49 and 2075–99 (except in June, where during 2025–49, no trend can be seen) and negative trends in the middle period are observed. However, opposite is the case for October and November (negative trends for the end periods and positive trends for the middle period). For July, rainfall first increases during 2025–74 and decreases during 2075–99. February (2050–99) and April (2025–99) show continuous increasing trends, and March (2025–99) shows continuous decreasing trends of rainfall. In case of MRICGCM3, rainfall shows continuous decreasing trends from 2025 to 2099 for April, May and September. Increase in trends is seen for March (2025–99), July (2050–99) and November (2050–99). For June, rainfall first increases during 2050–74 and then increases again in the next period. For October, rainfall first decreases from 2025–75 and then increases during 2075–99.For BCC-CSM1, rainfall shows continuous decreasing trends from 2025–99 for May, June, August and October. However, for March, April, September and November after showing decreasing trend during 2025–75 (no trend during 2025–49 for November), rainfall shows increasing trends during 2075–99. For the month of July, decreasing trends are observed during 2025–74 and increasing trend toward the end of the century. For January, February and December, rainfall remains same from 2050 to 99.

Under RCP8.5, for ACCESS, no trend is seen for January, February, November and December. Also, continuous decreasing trends in March and April and increasing trends in June and September are observed. For May and August, increasing trends during 2024–49 and 2075–99 and decreasing trend during 2050–74 are observed. Positive trends from 2050–99 and negative trends in 2025–49 are observed for July. For October, rainfall shows negative trends from 2025–74 and then positive trend for the period 2075–99. In case of CCSM4, for February and March, the rainfall trends are positive for the first two time periods and negative toward the end of the century. Negative trends are observed from 2050 to 99 for April and June. For July and October, trends become positive from 2050–99. Positive trends in the first and last periods and negative in the middle time period are seen for May and September. However, opposite is the case for August, where negative trends at the end periods and positive in the middle period are observed. For MIROC, positive trends from 2050 to 99 are observed for February, April, June and November. For July and October, positive, negative and then again positive trends are seen for the tree time periods. Continuous negative trends are observed only for the month of September. Negative trends in the first two periods become positive toward the end of the century for March. For May and August, there is a possibility that rainfall will increase for 2025–74 and will decrease during 2075–99. No change in trend is seen from 2025 to 74 in case of January and December which remains unchanged for December during 2075–99, while increasing trend is seen for January during this period. In case of MRICGCM3, for the months of February, March, July and October, rainfall shows decreasing trends from 2050 to 99. However, for January and December, no significant trend can be observed. Rainfall shows increasing trends from the middle of the century for April and November. Similar trends can be seen for June, August and September, where trends are positive for the first two time periods and negative again during 2075–99. However, for May, negative trends are seen for the first and last time period and positive for 2050–74. For the model BCC-CSM1, February and March show positive trends from 2025 to 74 and negative trend for 2075–99. No trend is observed for January and December. Alternate trends of positive, negative and positive are seen for April and July, and opposite alternate trends of negative, positive and negative are observed for August for the three time periods. For May and September, the trends are negative from 2025 to 99. For May and September, trends are negative from 2025 to 74 and positive again for 2075–99. For November, though there is no trend for the first and the last periods, it becomes positive in the middle period.

It is clear from the above results that one single model cannot exactly predict the future rainfall over the study basin. Nature of trend for the same month is different for different models. However, for almost every model, no significant trend is observed for January and December during the entire future period. It has been observed that, rainfall under RCP8.5 for almost every month will be more than that under RCP4.5 and the historical period. However, for some particular months, rainfall under RCP4.5 is less than the historical period depicting decrease of rainfall in those months compared to 1970–2005. As for example, rainfall during 2050–75 in case of BCC-CSM1.1 model under RCP4.5 is less than that of historical period. For the monsoon period, i.e., from May to August, if we compare the results, most of them show negative trends except ACCESS, indicating that there is a possibility of decrease in rainfall during the monsoon period in future.

4 Conclusions

The present study statistically analyzed the possible changes in rainfall trends in future over the Puthimari basin due to climate change using Mann–Kendall non-parametric test and the resultant z statistics. To carry out the study, five CMIP5 models have been considered under RCP4.5 and RCP8.5. The future rainfall up to 2099 has been analyzed considering three different time periods, and it was compared with the historical rainfall pattern. Different models gave different results because of which a single model cannot be applicable in prediction of rainfall. However, for almost all the models, trends for the months of January and December are insignificant. Rainfall under RCP8.5 for almost all the months is more than that under RCP4.5 and the historical period. Also, there is a likelihood of decrease in rainfall in the monsoon season from the middle toward the end of the century.

Conflict of Interest. None.

References

1. Taher M (1975) Regional basis of agricultural planning in the Brahmaputra Valley. J North East India Geogr Soc 7(1 & 2):9–18

2. Guilbert J (2016) The Impacts of climate change on precipitation and hydrology in the Northeastern United States, Graduate College Dissertations and Theses, Paper 646, University of Vermont
3. Knutti R, Sedlavcek J (2012) Robustness and uncertainties in the new CMIP5climate model projections. Nature Clim Change
4. Meher JK, Das L, Akhter J, Benestad RE, Mezghani A (2017) Performance of CMIP3 and CMIP5 GCMs to simulate observed rainfall characteristics over the western Himalayan region. J Clim 30:7777–7799
5. Sperber KR, Annamalai H, Kang IS, Kitoh A, Moise A, Turner A, Wang B, Zhou T (2013) The Asian summermonsoon: an intercomparison of CMIP5 vs. CMIP3 simulations of the late 20th century. Clim Dynam 41:2711–2744
6. Sun Q, Miao C, Duan Q (2015) Comparative analysis of CMIP3 and CMIP5 global climate models for simulatingthe daily mean, maximum, and minimum temperatures and daily precipitation over China. J Geophys Res Atmos 120:4806–4824
7. Taylor KE, Stouffer RJ, Meehl GA (2012) An overview of CMIP5 and the experiment design. Am Meteorol Soc 485–498
8. Moss RH, Edmonds JA, Hibbard KA, Manning MR, Rose SK, van Vuuren DP, Kram T (2010) The next generation of scenarios for climate change research and assessment. Nature 463(7282):747–756
9. Demirel MC, Moradkhani H (2016) Assessing the impact of CMIP5 climatemulti-modeling on estimating the precipitation seasonality and timing. Clim Change 135(2):357–372
10. Wayne GP (2013) The beginner's guide to representative concentration pathways (RCPs), 1.0, pp 1–24, Skeptical Science. https://www.skepticalscience.com/docs/RCP_Guide.pdf
11. Arnell NW (1999) Climate change and global water resources. Glob Env Change 9:31–49
12. Houghton JT, Ding Y, Griggs DJ (2001) Climate change 2001—the scientific basis contribution of Working Group I to the third assessment report of the intergovernmental panel on climate change (IPCC). Cambridge University Press, London, p 944
13. Stocker TF, Clarke GKC, Le Treut H, Lindzen RS, Meleshko VP, Mugara RK, Palmer TN, Pierrehumbert RT, Sellers PJ, Trenberth KE, Willebrand J (2004) Physical climate processes and feedbacks. In: Houghton JT et al (eds) Climate change 2001: the scientific basis. Cambridge University Press, Cambridge, pp 417–470
14. Changchun X, Yaning C, Weihong L (2008) Potential impact of climate change on snow cover area in the Tarim River basin. Environ Geol 53:1465–1474
15. Watterson IG, Whetton PH (2011) Distributions of decadal means of temperature and precipitation change under global warming. J Geophys Res 116:D07101. https://doi.org/10.1029/2010JD014502
16. United Nations Framework Convention on Climate Change (UNFCCC) (2010) Climate change: impacts, vulnerabilities and adaptation in developing countries
17. World Bank Report (2013) Warming climate in India to pose significant risk to agriculture, water resources, health (The World Bank, 19 June 2013)
18. Postel SL, Daily GC, Ehrlich PR (1996) Human appropriation of renewable fresh water. Science 271:785–788
19. Patowary AN, Goswami K, Hazarika P, Pathak B (2017) A study on long term rainfall pattern of Dhubri and Guwahati in Assam, India: a Time Series Approach. Int. J. advanced research in computer science, 8(7), 708–713.
20. Rehman N, Adnan M, Ali S (2018) Assessment of CMIP5 climate models over South Asia and climate changeprojections over Pakistan under representative concentration pathways. Int J Glob Warming 16(4):381–415
21. Rupp, D.E., Abatzoglou, J., Hegewisch, K., and Mote, P.(2013). Evaluation of CMIP5 20th century climate simulations for the Pacific Northwest US. J Geophys Res Atmos 118(19):10884–10906

Chapter 4
Impact of Sea-Level Rise on a Coastal Catchment of Brunei Darussalam

U. Ratnayake(✉), S. N. Amirah Arun, E. K. Abdul Rahman, and S. Shams

Faculty of Engineering, Universiti Teknologi Brunei, Gadong 1410, BE, Brunei Darussalam
uditha.ratnayake@utb.edu.bn

Abstract. This study presents an impact assessment of climate change in the Brunei River catchment that leads to the occurrences of tidal flood from sea-level rise. Brunei Darussalam is a small country located along the coast of Borneo Island on the South China Sea. Case studies of several countries for flood hazard are taken into consideration to understand flood simulation models and predictions of future flooding event after calibration and validation. The impact of tidal flood for Brunei River catchment is predicted using HEC-RAS software. A 2D flood model of Brunei River catchment is formulated to simulate the sea-level rise due to tidal flood in the year 2050 and 2100 by using a Rapid Ice Melt (RIM) and IPCC-AR5 projections. Simulations show that most populated inland flood plains of the country may become susceptible to tidal floods by 2100. However, the results of tidal flood maps are highly dependable on the accuracy of the digital elevation model as well as the predictions. This model considers only the internal tidal inundation and not the inundations in the coastal areas.

Keywords: Flood modelling · 2D flood simulation · HEC-RAS · Sea-level rise · Climate change

1 Introduction

Climate change is a current phenomenon causing many adverse impacts all over the earth. Chaotic weather, increased rainfall or droughts, increased temperatures, glacial melts, forest fires, ecological threats and health problems are some of such impacts where this study concentrates on sea-level rise. A primary cause of climate change is global warming due to atmospheric accumulation of greenhouse gases. [10] have suggested that even under the assumption of large amounts of negative carbon dioxide emissions, the thermosteric sea-level rise is not reversible for at least several centuries. Therefore, this sea-level rise hazards need to be estimated, so that appropriate and feasible preventive and adaptive measures can be considered. Simulation of coastal environments for tidal flooding of the low-lying areas will portray the more frequent adverse impacts on the inhabitants.

R. K. Bhattacharjya et al. (eds.), *Sustainable Water Resources Management*,
Advances in Sustainability Science and Technology,
https://doi.org/10.1007/978-981-16-7535-5_4

The capital city of Brunei Darussalam, Bandar Seri Begawan and its suburbs are located on low-lying plains of Brunei River. A significant part of the Brunei River catchment is affected by the tidal influences of the Brunei Bay. Hasan et al. [4] have suggested that climate change in recent years has been affecting the country by increased rainfall and temperature. Because of this reason, it is essential to predict the tidal inundation in Brunei River catchment, so that possible damages and remedial measures could be studied and addressed earlier. This study aims to simulate the flooding due to sea-level rise in Brunei River catchment using a 2D flood model. The flood model is based on HEC-RAS software, and it is calibrated for the past flood events and then used to model the future predictions of sea-level change. The outcome of the study is to demarcate inundation areas in the Brunei River catchment using various future sea-level change predictions.

2 Impact of Climate Change on Flooding

Climate patterns play a fundamental role in shaping natural ecosystems, economies and cultures directly or indirectly. Arnell and Gosling [1] have shown that the impacts of climate change on river flood frequency characteristics have consistently increased the flood magnitude across humid tropical Africa, South and East Asia, South America, high latitudes of Asia and North America. They showed that with the HadCM3 climate model and data for SRES A1b scenario, the current 100-year flood would occur twice as often across 40% of the world and over 60% of South East Asia, Central Africa, Eastern Europe and Canada. Hoppe et al. [5] have indicated that the local governments should allocate significant budgets, time and effort to formulate and implement local climate change policies that give enough attention to both mitigation and adaptation.

Shaviraachin [9] successfully used 2D overland and channel flow model of SOBEK to generate a flood map for the basin to provide relevant information on the behaviour of flood and its effect on settlements. The author has shown that Mozambique, especially the Limpopo basin, is highly vulnerable to a weather-related natural disaster like drought and floods.

The discretisation of the land area is a factor that affects both the results and simulation performance. Hsu et al. [6] have used a newly developed hydrodynamic computing program 3Di to investigate the influence of DEM resolution for the flood inundation simulation. The result shows that the inundation area evaluation increases with coarser DEMs, whereby the 40 × 40 m DEM gave an inundation area of 1.5 times greater than 1 × 1 m DEM. Patel et al. [8] have assessed floods of Surat city of India, which is located at 100 km downstream of Ukai Dam. Surat city has experienced the most significant flood in the year 2006 due to the peak discharge of about 25,770 m^3/s released from Ukai Dam. The assigned Manning's roughness values for agriculture (0.07), the build-up (0.2), forest (0.035), grassland and grazing land (0.045), wastelands (0.025) and wetlands (0.12) have been used in the model and validated for the flood event in August of the year 2006. The results have shown that when simulated under the bank protection work, the major zones of the city are safe against inundation up to 14,430 m^3/s—except for the west zone.

Fu et al. [3] have carried out a study to understand the flood–tide interaction of Min River for the Minjiang Estuary, and a high-resolution flood–tide coupled model with GIS data for the Minjiang Estuary is used for the study. An unstructured triangular mesh is used to portray complex topography and the high-resolution grids to portray the topography of 50–100 m at key areas accurately. The result has shown that the simulation result of each tide gauge agrees very well with the measured data. Stronger flood signal and weaker tidal signal have shown farther away from the Minjiang Estuary and vice versa when it is closer. The velocity has reduced during high water tide and has increased during low water tide. The reason for this is due to the opposite flow direction of the tide and flood, as the flood flow direction is seaward side along the river channel.

3 Development of the Simulation Model

Drainage system consisting of the man-made channels and natural streams of the Brunei River has been reconstructed in the HEC-RAS software using more than 2500 cross sections. These cross sections are then converted to a channel surface model and etched on to the DEM of the catchment. The entire catchment has defined as the 2D simulation area as most of the catchment is seen to have influenced by the tidal flows. The land use of the catchment is categorised as built-up areas, forests and streams to assign the roughness values.

The initial water in the channel system is set to mean tidal water level, and no upstream water flows are introduced to the system. This simulation under the influence of the sea-level rise and the tidal effects considers only the downstream boundary conditions. The downstream boundary condition at the river mouth is set, as indicated in Table 1, and the maximum 2016 tide has the river stage variation, as shown in Fig. 1. It is assumed here that the tidal variations will not be changed due to climate change. Two different sea-level rise prediction approaches, i.e. the sea-level rise results from this Rapid Ice Melting (RIM) and the results from the Fifth Assessment Report of the Intergovernmental Panel on Climate Change (IPCC-AR5) [7] are used in this study. The RIM approach is adapted from Cynthia et al. [2], whereby the possible future floods for New York City is prepared.

Table 1 Designed experiments to evaluate tidal impacts of the Brunei River

Experiment	Sea level (m asl)	Tidal variation
Current conditions	0.00	Maximum 2016
IPCC-AR5 prediction 2050	0.30	Maximum 2016
RIM prediction 2050	0.66	Maximum 2016
IPCC-AR5 prediction 2100	0.81	Maximum 2016
RIM prediction 2100	2.03	Maximum 2016

4 Results and Discussion

The simulation of the Brunei River catchment under the tidal effects shows the dynamic response of the Brunei River and its drainage system. Observations on the river stage

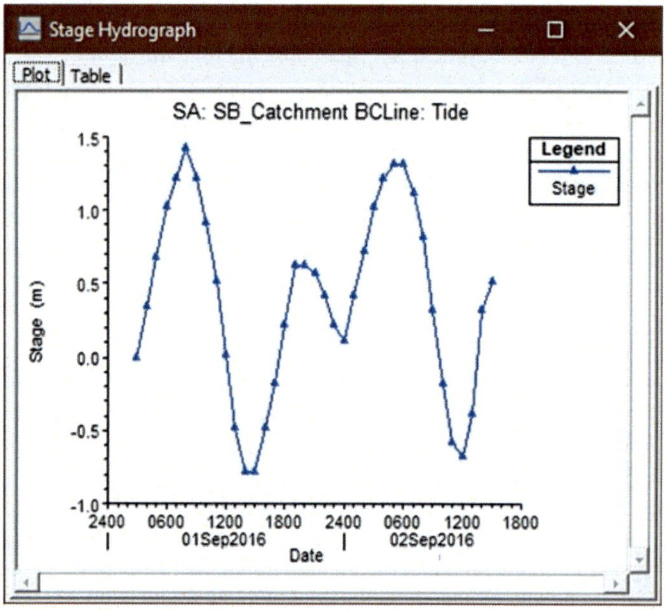

Fig. 1 Stage variation at the river outfall

or the flow are not gauged in the Brunei River due to its high sensitivity to tidal surges. Therefore, the calibration of the model is very challenging. However, observed inundation extents through variation of the roughness values are used for the calibration of the model. The river and drainage system followed recommended values, and the overland roughness is changed to accommodate observed inundations.

Figure 2 shows the inundation due to 2016 highest tide. These correspond to the maximum depths reached at any location during the considered period. The dynamic variations of the inundation depths and extents are used to calibrate the model by comparing them with field data. Although the model output and the observation data are comparable, calibrations could be improved in future by incorporating the flow velocity measurements. Present maximum inundation within the Brunei River catchments as in Fig. 2 shows that the flooding is limited to the mangroves and low-lying forest areas adjacent to the river. As such, no impact on the livelihood is seen. However, it appears that in certain areas water levels became closer to the overflow levels. Both model output and field observation indicate that the river and most of the connecting streams are affected by the tide.

According to the IPCC-AR5 (2013), the sea-level rise is projected for RCP2.6, RCP4.5, RCP6.0 and RCP8.5 scenarios which are 28–61, 36–71, 38–73 and 52–98 cm, respectively. Leaving aside the unlikely scenario of the RCP2.6, the maximum values of the ranges are averaged to get the future applicable sea level by 2100. This resulted in 30 cm of sea-level rise by 2050 and 81 cm of sea-level rise by 2100. This sea-level rise produced a little higher inundation, as shown in Fig. 3, compared to the current maximum high tide inundation. Even though the whole river and major connected streams are affected, the overflowing of water to the land is not observed in most places except for

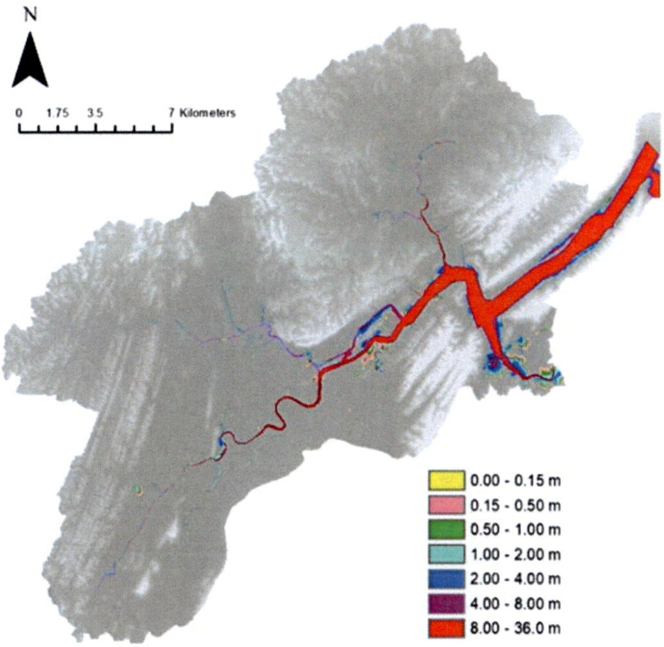

Fig. 2 Inundation from maximum tide in 2016

a very few locations. It is found that the areas which were overflowing have submerged grounds that are still within the mangroves and forested areas. As shown in Table 2, the total area of inundation including the river and streams increased from 26.78 × 10^6 to 28.96 × 10^6 m^2. Therefore, for IPCC-AR5 prediction for 2050, an additional 2.18 × 10^6 m^2 is inundated. Also, the land inundations ranged from 0.0 to 4.0 m increased significantly.

The IPCC-AR5 year 2100 prediction of 0.81 m sea-level rise caused inundation of nearly all the low-lying lands as seen in Fig. 4, which includes the most populated as well as commercial and industrial areas. The total area of inundation given in Table 2 increased from 26.78 × 10^6 to 79.33 × 10^6 m^2. In this submerged area, 0.00–0.15 m depth inundation covered 33.58 × 10^6 m^2 area while 0.150–0.50 m depth inundation covered 16.15 × 10^6 m^2. In other words, the additional inundations of 2100 compared to 2050 are 32.83 × 10 m and 14.60 × 10 m, respectively and such inundation will be a significant threat to the economy (Figs. 5 and 6).

By using RIM prediction approach [2], the rise of sea level for the year 2050 and 2100 are 0.66 m and 2.03 m, respectively. These values are much higher than the values proposed by IPCC-AR5 prediction, whereby the total inundations increased to 56.62 × 10^6 m^2 and 97.94 × 10^6 m^2, respectively for the RIM 2050 and RIM 2100. The RIM 2050 prediction increased the inundation area of the depth 0.00–0.15 m to 25.39 × 10^6 m^2 while the RIM 2100 predicts the depth of 1.00–2.00 m to be about 57.07 × 10^6 m^2 area. Table 2 values show an increase in flood depth rather than a spread in the flood area indicates that the whole flood plain has submerged.

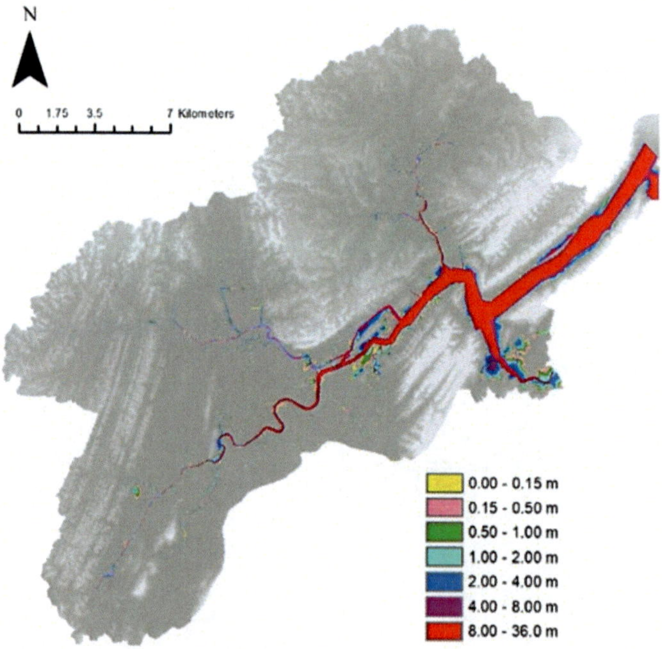

Fig. 3 IPCC-AR5 sea-level rise by 2050 (0.30 m)

Table 2 Inundated land area (10^6 m^2) for each prediction

Inundation depth (m)	Inundated area (10^6 m^2)				
	Historical	AR2050	AR2100	RIM2050	RIM2100
0.0–0.15	0.75	1.06	33.58	25.39	2.05
0.15–0.50	1.54	1.93	16.15	3.36	4.40
0.50–1.00	1.72	2.12	3.35	2.64	6.78
1.00–2.00	2.44	2.88	3.90	3.40	57.07
2.00–4.00	2.88	3.18	3.95	3.60	7.22
4.00–8.00	3.74	3.79	4.05	3.98	5.04
8.00-max	13.67	13.96	14.33	14.22	15.34
Total	26.77	28.96	79.33	56.62	97.94

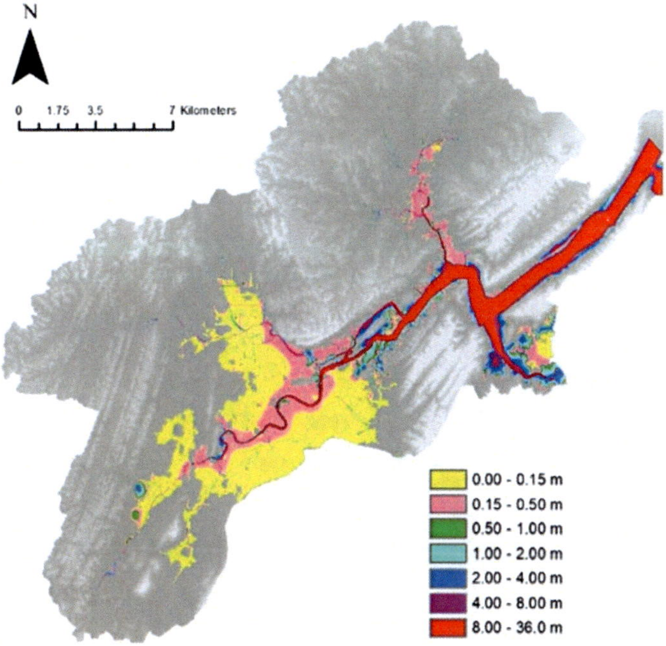

Fig. 4 IPCC-AR5 sea-level rise by 2100 (0.81 m)

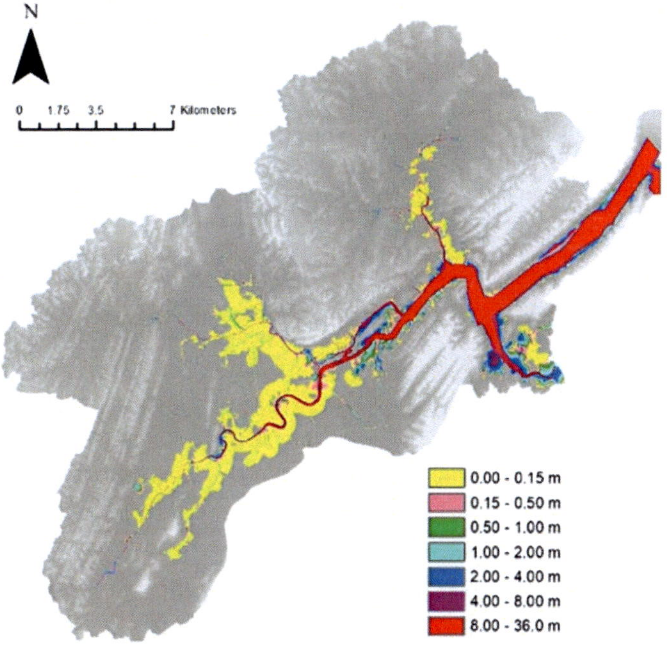

Fig. 5 RIM sea-level rise by 2050 (0.66 m)

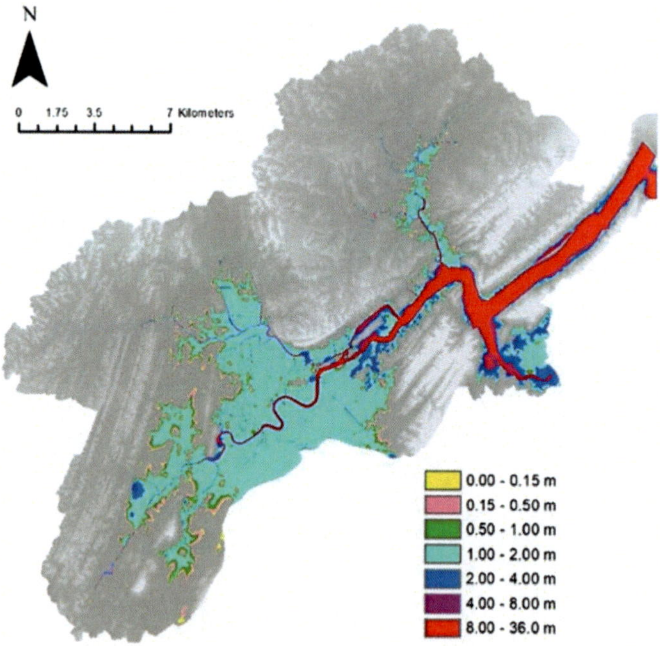

Fig. 6 RIM sea-level rise by 2100 (2.03 m)

Figure 7 shows the time response of the catchments at various locations. Graph 'a' shows the tidal variation close to the river mouth, and it is very much equivalent to the boundary condition of tidal variation plus the raised sea level. The graphs 'd', 'e', 'f', 'g' and 'h' show the water level in the channels where the tidal inflow and subsequent draining out can be observed. The result indicates that the velocities are high in the channels, which indicate that the current channels are adequate to drain out the floodwater quickly from the area. However, the graphs 'b' and 'c' show the overland flow velocity is very low, there is high accumulation of flood water, and this is because the areas are covered with forest with a less dense network of the drainage system.

5 Conclusion

Climate change causes many adverse impacts, and the rise of sea level is one such impact affecting globally as one-third of the world population is living in coastal areas. Brunei Darussalam is one such country where its capital and the highest populated district is located in a tidal river catchment. The tidal floods into the flood plains are modelled using HEC-RAS considering the sea-level change up to 2100. Two prediction approaches, IPCC-AR5 and Rapid Ice Melting (RIM), are used for the study. The study suggests that by using IPCC-AR5 predictions, the impact by 2050 is not significant. However, the RIM prediction indicates that in 2050 flooding may occur in the main commercial district and some residential areas. By 2100, both projections show that nearly all the flood plain of the Brunei River may be inundated by sea-level rise and tides, which may

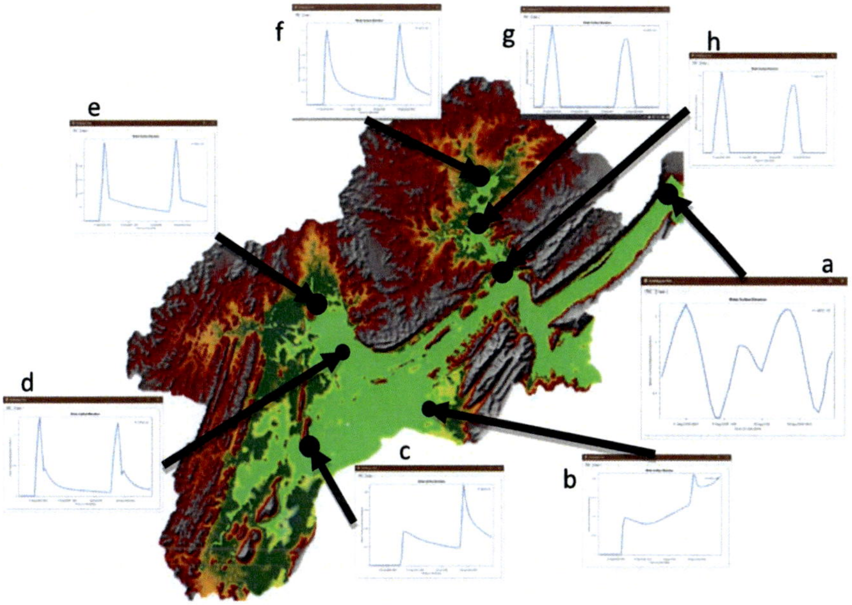

Fig. 7 Time variation of flood flow under IPCC-AR5 2100 prediction

affect almost all the low-lying residential and commercial areas. These predicted results suggest that further study is required to look at mitigatory measures of the tidal floods and floods due to rainfall.

References

1. Arnell NW, Gosling SN (2016) The impacts of climate change on river flood risk at the global scale. Clim Change 134(3):387–401. https://doi.org/10.1007/s10584-014-1084-5
2. Cynthia R, Vivien G, Radley H, Daniel B, Richard G (2011) More floods ahead: adapting to sea level rise in New York City [ONLINE]. https://www.giss.nasa.gov/research/briefs/rosenzweig_03/. Accessed 8 Mar 2020
3. Fu C et al (2017) Flood-tide interaction numerical simulation at Min River tidal reach. Procedia IUTAM 25:119–125. https://doi.org/10.1016/j.piutam.2017.09.018.40 (Elsevier B.V)
4. Hasan Ali SNA, Ratnayake U, Shams S, Nayan ZH, Rahman EKA (2017) Analysis of temperature and precipitation trends in Brunei Darussalam using statistical downscaling model. J Theor Appl Climatol. https://doi.org/10.1007/s00704-017-2172-z
5. Hoppe T, van den Berg MM, Coenen FH (2014) Reflections on the uptake of climate change policies by local governments: facing the challenges of mitigation and adaptation. Energy Sustain Soc 4(1):8. https://doi.org/10.1186/2192-0567-4-8
6. Hsu YC et al (2016) An investigation of DEM resolution influence on flood inundation simulation. Procedia Eng 154:826–834. https://doi.org/10.1016/j.proeng.2016.07.435 The Author(s)
7. IPCC-AR5 (2013) The physical science basis. Contribution of working group I to the fifth assessment report of the intergovernmental panel on climate change [Stocker TF, Qin D, Plattner G-K, Tignor M, Allen SK, Boschung J, Nauels A, Xia Y, Bex V, Midgley PM (eds)].

Cambridge University Press, Cambridge, United Kingdom and New York, NY, USA, 1535 pp

8. Patel DP et al (2017) Assessment of flood inundation mapping of Surat city by coupled 1D/2D hydrodynamic modelling: a case application of the new HEC-RAS 5. Nat Hazards 89(1):93–130. https://doi.org/10.1007/s11069-017-2956-6
9. Shaviraachin T (2005) Flood simulation—a case study in the Lower Limpopo Valley, Mozambique using the SOBEK flood model. MSc thesis, International Institute for Geo-Information Science and Earth Observation, The Netherlands
10. Tokarska KB, Zickfeld K (2015) The effectiveness of net negative carbon dioxide emissions in reversing anthropogenic climate change. Environ Res Lett 10:094013. https://doi.org/10.1088/1748-9326/10/9/094013

Chapter 5
Challenges of Data Scarcity in Statistical Downscaling of Rainfall Using Large-Scale GCM Models

Jayshree Hazarika[1(✉)] and Arup Kumar Sarma[2]

[1] Civil Engineering Department, Assam Engineering College, Jalukbari, Guwahati, Assam 781013, India
jayshree@iitg.ac.in

[2] Civil Engineering Department, Indian Institute of Technology Guwahati, Guwahati, Assam 781039, India
aks@iitg.ac.in

Abstract. In climate change impact studies, statistical downscaling is considered to be quite effective while predicting future data of any meteorological parameter. However, complications arise when available data is of short duration. In the northeastern states of India, researchers have been facing this difficulty of scarcity of data for a long time. Short duration of observed data makes it difficult to choose suitable datasets for calibration and validation. The future dataset obtained from downscaling these datasets will be acceptable, if the calibrated model is precisely validated before using in future prediction. Therefore, we need a new method to find out a suitable dataset for calibrating the model. In the present study, hence, a novel approach is being tried to determine the best fit dataset for calibrating a statistical downscaling model. Five IMD stations, situated in the northeastern region are considered as focused area in this study. Six different combinations of calibration–validation sets are compared to find out their effectiveness and related challenges. To find out the correlation of the calibration–validation datasets, statistical parameters are evaluated for all the six combinations, and for all the stations. The results suggest that, the best combination is the one with higher R^2 value and lower errors in RMSE, RE_μ, RE_б and percentage change in maximum. Hence, we may say that lower the errors, better is the combination and better is the model calibrated.

Keywords: Statistical downscaling · Calibration and validation · GCM

1 Introduction

Climate change impacts on rainfall characteristics have large influence on the development of a region. Extensive study on rainfall characteristics taking climate change impacts into account has gained vast attention from the research community. IPCC's

R. K. Bhattacharjya et al. (eds.), *Sustainable Water Resources Management*,
Advances in Sustainability Science and Technology,
https://doi.org/10.1007/978-981-16-7535-5_5

Fourth Assessment Report (AR4) explains that in the last 100 years global average temperature has increased about 0.74 °C, two-third of which has increased during the last three decades. A number of studies on hydrology and climate change have been done with emphasis on the Indian subcontinent. Major impacts of climate change in India would be on hydrology, water resources and agriculture [4]. Since GCM variables are commonly used for climate studies, therefore, investigations are done on establishment of empirical relation between local and climatic variables and comparisons are made among various statistical downscaling methods using GCM output [9, 10]. Zhang [11] developed a statistical downscaling method of GCM output for prediction of soil erosion and crop production. Gosain et al. [2] developed a hydrologic modelling considering climate change effect for various river basins in India. There is a need to understand the influence of climate on hydrologic extremes, integrating the hydrological and atmospheric models [5]. Many research projects are initiated to study the climate change impacts on hydrology of Indian basins, with emphasis on the northeastern region. Climate change impacts are studied on the rainfall characteristics and streamflow behavior of Brahmaputra as well as Barak basin, with the use of downscaling methods [1, 6]. These studies indicate significant changes in the rainfall pattern of these basins. High intensity rainfall of short duration and longer dry spells will have adverse effect on flood and drought scenario of the region [7].

In these studies, the most crucial part is the scarcity of observed data. In most cases, 30 years of data is considered to be enough for calibrating a meteorological trend. Some researchers have also used short period of data (10 years). But there is no proper recommendations or guidelines that suggests what is the best data length to get a satisfactory result for any meteorological study. In northeastern states of India, data scarcity is a very common case both in temporal and spatial scale. The orographic variations in the region is very high. Compared to that, number of meteorological stations available is relatively small. Moreover, recorded datasets available from some of the stations are of short length, which might be because of less accessibility to the remote locations of those stations. When the observed data collected is less, it is difficult to choose which set of data should be used for calibration and which set should be used for validation. The future dataset obtained from downscaling will be acceptable, if the calibrated model is precisely validated before it is used for prediction. Previous studies show that data scarcity degrades the quality of downscaling results by overestimating or underestimating the downscaled parameters. Wagner et al. [8] stated that limited baseline period had led to underestimation of temperatures in Western Ghats, India. Also, many literatures can be found where researchers have faced this issue and hence tried new approaches to make the downscaling results better [3, 8, 12].

Therefore, we need a new method for finding out the best fit dataset for calibrating the model. Keeping this point in mind, in the present study, different combinations of calibration–validation sets are compared to find out their effectiveness and related challenges. Based on the results, an attempt is made to find out the best combination of data for calibration as well as validation that will precisely predict the future pattern of rainfall in data scares areas.

2 Materials and Methods

Data Used

Two types of data have been used in this study, namely

(i) Observed rainfall data
    Observed daily rainfall data of each station has been collected from Indian Meteorological Department (IMD) under an MOU between IIT Guwahati and IMD. The time period of data collection is from 01-01-1969 to 31-01-2012. However, no data are available for any station from 2001 to 2005.
(ii) GCM (Global Climate models) data
    HadCM3 (Hadley Centre Coupled Model, Version 3), is developed by Hadley Centre in the United Kingdom, under Intergovernmental Panel on Climate Change's (IPCC) Fourth Assessment Report (AR4). The emission scenario considered in the present study for analysis is A2, which is described in IPCC's Special Report on Emission Scenarios (SRES) as a scenario representing a very heterogeneous world with continuously increasing global population and regionally oriented economic growth. HadCM3 monthly data are available from 2000 to 2100.

Study Area

Five IMD stations, situated in the northeastern region are considered as focused areas in this study (Fig. 1). All these stations are located in one of the most important cities/towns in the northeastern region. Geographic details of the stations are provided in Table 1.

Downscaling Using Stepwise Multiple Linear Regression (SMLR)

Downscaling of rainfall has been done at the five IMD stations (Fig. 1), with the use of multiple linear regression.

In multiple linear regressions, the procedure is divided into four basic steps, namely: specification, calibration, validation and forecast. In the specification stage, the model and correlated predictors are selected. In calibration stage, the relation between the output data (predictand) and input data (predictors) is attained. The precision of the model is checked in the validation stage, while in the forecast stage, validated model is used to predict the future dataset.

Assumptions

The following assumptions are considered while using the multiple linear regression:

(i) The relation between Y (predictand) and $X_1, X_2, \ldots, X_n$ (predictors) are linear.
(ii) The residuals have a constant variance (σ) and are normally distributed.
(iii) There is no autocorrelation among the predictors.
(iv) X variables are fixed.

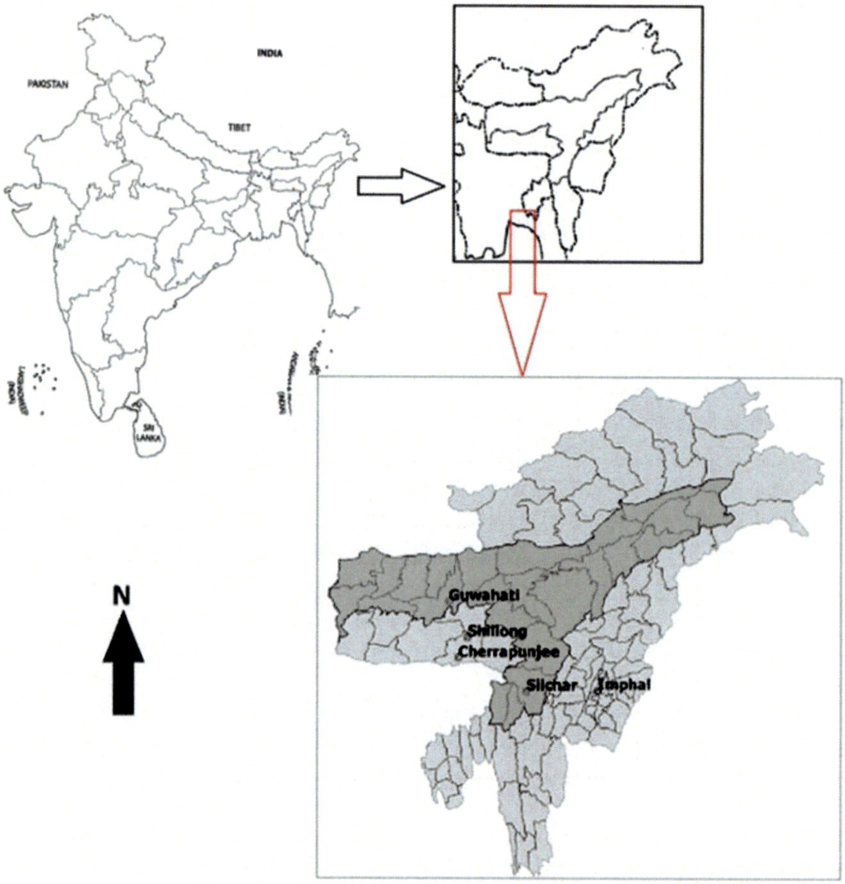

Fig. 1 Geographical location of the five IMD stations in the study area

Table 1 Geographic details of the stations considered for the study

Sl. No	Name of the stations	Region	Latitude	Longitude	Elevation
1	Cherrapunjee	Meghalaya	25° 15′ 0″ N (25.25 °)	91° 44′ 0″ E (91.7333 °)	1300 m (4265 ft)
2	Imphal	Manipur	24° 45′ 36″ N (24.7599 °)	93° 53′ 48″ E (93.8967 °)	774 m (2539 ft)
3	Shillong	Meghalaya	25° 34′ 0″ N (25.5667 °)	91° 53′ 0″ E (91.8833 °)	1600 m (5249 ft)
4	Silchar	Assam	24° 45′ 0″ N (24.75 °)	92° 48′ 0″ E (92.8 °)	21 m (70 ft)
5	Guwahati	Assam	26° 6′ 22″ N (26.1061°)	91° 35′ 9″ E (91.5859°)	54 m (177 ft)

Standardization

Before calibration, the large-scale climate variables need to be processed. In the process of standardization, the mean (μ) is subtracted from the ith predictor/predictant, and then, it is divided by the standard deviation.

$$V_{\text{std}}(n) = \frac{v_i(n) - \mu(n)}{\sigma(n)} \tag{1}$$

where V_{std} is the standardized data of nth predictor.

v_i is the ith variable of the nth predictor
μ is the mean of all the variables of nth predictor, and
σ is the standard deviation.

Selection of Predictors

The predictors are selected by step wise regression method. This method consists of two main approaches, forward selection and backward elimination. The coefficient of determination is obtained between the observed data and a particular predictor. The coefficient of determination lies between 0 and 1. Larger the value, stronger will be the correlation.

Correlation

It is the statistical relation between the arrays of two variables. It shows how they are correlated. Pearson correlation is mostly used to determine the coefficient of correlation. Its value lies between –1 to +1. Larger the absolute value of the correlation, stronger will be the relationship between the two variables.

Model Calibration and Validation

There are several ways for performing calibration. Some of these are:

(i) Multiple linear regressions without additive constant

$$y_i = \beta_1 x_{i1} + \beta_2 x_{i2} + \cdots + \beta_n x_{in} \tag{2}$$

(ii) Multiple linear regressions with additive constant

$$y_i = \beta_1 x_{i1} + \beta_2 x_{i2} + \cdots + \beta_n x_{in} + r_i \tag{3}$$

(iii) Multiple linear regression with a multiplying constant

$$y_i = (\beta_1 x_{i1} + \beta_2 x_{i2} + \cdots + \beta_n x_{in})m \tag{4}$$

Here, y_i—rainfall, β—coefficient, x_i—predictor, r_i—residual, m_i—multiplying constant.

Future Forecast

In this step, the relationship attained from calibration between the predictand and the predictors, that was tested for precision in validation, is used to predict the future data of the predictand with the use of future data series of the predictors.

3 Results and Discussion

Screening of Predictors

Screening of predictors has been done as per the method explained for SMLR in the previous section. There are 25 predictors available for HadCM3 model. It is not very wise to use all the available predictors in the SMLR for calibration, which will make the process tedious. Hence, an initial screening has been done by observing the correlation of these predictors with the observed data. Those predictors having corrections well enough to be used for calibration are considered. For this purpose, Pearson correlations between total monthly rainfall and predictors are calculated for each of the 5 stations. The list of selected predictors is given in Table 3. The correlations were calculated for each combination separately, which is explained in the next paragraph.

Calibration and Validation

The base period considered for calibration and validation is very short, since only six years data are available (2006–2011). Here, it is required to be mentioned that, for the year 2012, only January month's rainfall data was available. Hence, data for January' 2012 was excluded from the study. In the present study, different combinations are tried to find out the best combination of data for calibration. Six combinations are considered here as tabulated below in Table 2. SMLR has been performed for all the six combinations separately to find out the final set of predictors, which are different for different combinations. Selected predictors and their R^2 values (co-efficient of determination) are tabulated below in Table 3. From the R^2 values (Table 3), it is observed that combination 6 (5 years for calibration and 1 year for validation) is giving the best results in case of all the stations. But, some might say that this may be due to the short period of data set considered for validation. Hence, to find out the correlation of the calibration–validation datasets, statistical parameters are evaluated for all the six combinations, and for all the stations. The parameters considered are, Pearson correlation, coefficient of determination (R^2), root mean square error (RMSE), relative error in mean (RE_μ), relative error in standard deviation (RE_б) and percentage change in the maximum value of the series. These statistical parameters results are tabulated below (Table 4).

From the above statistical parameter analysis, it is observed that in most of the cases the similarities between the two datasets (calibration–validation) are lowest for combination 6, yet this combination is giving better results than the other combinations. This can be attributed to the reason that the calibration period is longer than the validation period, which makes the calibrated model more reliable due to consideration of more points (or data from the observed series) in calibration. However, it should be noted that the correlation coefficients (Person correlation and coefficient of determination, R^2)

Table 2 Different combinations of base period for calibration

	Combination 1		Combination 2		Combination 3	
	Calib	Valid	Calib	Valid	Calib	Valid
Years	2006–2007–2008	2009–2010–2011	2009–2010–2011	2006–2007–2008	2006–2008–2010	2007–2009–2011
	Combination 4		Combination 5		Combination 6	
	Calib	Valid	Calib	Valid	Calib	Valid
Years	2007–2009–2011	2006–2008–2010	2006–2007–2008–2009	2010–2011	2006–2007–2008–2009–2010	2011

Note calib: calibration period; valid: validation period

Table 3 Selected predictors for calibration and their R^2 values (Total Monthly rainfall)

Name of the stations		Combination 1		Combination 2		Combination 3		Combination 4		Combination 5		Combination 6	
		Calib	Valid	Calib	Valid	Calib	Valid	Calib	Valid	Calib	Valid	Calib	Valid
Cherrapunjee	R square	0.62	0.60	0.65	0.58	0.63	0.53	0.70	0.42	0.64	0.55	0.63	0.70
	predictors	hur200		psl		hur200		hur500		hur200		hur200	
		ta200		ta500		hur500		psl		psl		psl	
		Ts		zg200		psl		ta200		ta200		tas	
		zg200		prc		ta200		zg200		zg200		ts	
						ta500							
Imphal	R square	0.68	0.52	0.75	0.53	0.73	0.60	0.70	0.47	0.66	0.64	0.64	0.74
	predictors	hur850		pr		pr		pr		hur850		pr	
		zg200		prc		prc		prc		ta500		prc	
		zg500		ua500		ua500		ta500		zg500		zg500	
		zg850		zg850		uas		ua200		zg850		zg850	
						hur500							
Shillong	R square	0.64	0.65	0.71	0.51	0.72	0.52	0.73	0.34	0.68	0.60	0.67	0.62
	predictors	hur500		hur500		hur500		pr		hur500		ta200	
		ta200		hur850		pr		prc		ta200		ta850	
		ta850		ta850		ua200		ua200		ta850		hur850	
		zg850		tas		ts		zg500					

(*continued*)

Table 3 *(continued)*

Name of the stations		Combination 1		Combination 2		Combination 3		Combination 4		Combination 5		Combination 6	
		Calib	Valid	Calib	Valid	Calib	Valid	Calib	Valid	Calib	Valid	Calib	Valid
				zg850		hur200							
Silchar	R square	0.75	0.70	0.71	0.72	0.68	0.76	0.86	0.64	0.74	0.71	0.72	0.79
	predictors	hur200		pr		ta500		prc		hur200		prc	
		ta500		ta200		zg500		ta200		ta500		zg500	
		zg500		ua200		zg850		ua200		zg500		zg850	
		zg850		ta500				zg850		zg850		ta500	
Guwahati	R square	0.70	0.59	0.64	0.62	0.79	0.52	0.68	0.58	0.68	0.67	0.71	0.74
	predictors	ta500		tas		psl		hur500		hur500		psl	zg850
		zg500		zg200		ta500		psl		psl		ta500	
		zg850		ta500		zg500		tas		tas		tas	
		hur500				zg850		zg500		zg500		zg200	
								zg850		zg850		zg500	

Table 4 Statistical parameters of the combination sets (Total monthly rainfall)

Name of the stations	Statistical parameters	combination 1	combination 2	combination 3	combination 4	combination 5	combination 6
Cherrapunjee	Pearson	0.91	0.91	0.92	0.92	0.84	0.75
	R^2	0.84	0.84	0.85	0.85	0.70	0.57
	RMSE	400.53	400.53	367.05	367.05	503.36	639.07
	RE_μ	6.14	6.54	10.22	11.38	3.80	23.02
	RE_б	15.61	18.50	6.94	7.45	9.26	28.99
	%age ch in max	19.29	23.91	14.46	12.64	19.29	55.22
Imphal	Pearson	0.91	0.91	0.90	0.90	0.92	0.83
	R^2	0.82	0.82	0.81	0.81	0.85	0.69
	RMSE	36.46	36.46	35.93	35.93	50.83	75.73
	RE_μ	3.56	3.44	1.50	1.52	22.04	8.09
	RE_б	19.24	16.14	7.46	6.94	37.84	47.48
	%age ch in max	29.72	22.91	10.67	9.64	29.72	10.67
Shillong	Pearson	0.92	0.92	0.94	0.94	0.86	0.85
	R^2	0.85	0.85	0.88	0.88	0.74	0.72
	RMSE	68.56	68.56	81.34	81.34	95.39	109.05
	RE_μ	4.78	5.02	24.26	19.52	1.64	1.21
	RE_б	0.21	0.21	36.67	26.83	6.25	16.77
	%age ch in max	1.78	1.81	26.60	21.01	1.78	1.78

(*continued*)

Table 4 (*continued*)

Name of the stations	Statistical parameters	combination 1	combination 2	combination 3	combination 4	combination 5	combination 6
Silchar	Pearson	0.96	0.96	0.97	0.97	0.91	0.90
	R^2	0.92	0.92	0.95	0.95	0.82	0.57
	RMSE	74.87	74.87	62.52	62.52	121.49	102.82
	RE_μ	10.44	9.45	11.93	13.55	21.07	8.43
	RE_б	16.10	13.87	17.35	20.99	24.32	0.53
	%age ch in max	3.55	3.43	23.34	30.44	3.55	33.11
Guwahati	Pearson	0.94	0.94	0.82	0.82	0.95	0.80
	R^2	0.88	0.88	0.68	0.68	0.90	0.64
	RMSE	38.29	38.29	68.80	68.80	39.85	75.35
	RE_μ	1.94	1.90	12.13	13.80	5.16	27.62
	RE_б	18.38	15.53	9.43	10.42	18.29	18.25
	%age ch in max	43.57	30.35	17.14	20.68	20.68	43.61

should not be very low or exceptionally low. Furthermore, the other error parameters, namely root mean square error (RMSE), relative error in mean (RE_μ), relative error in standard deviation (RE_б) and percentage change in the maximum value of the series, should be within considerable range. Further observation of the statistical results shows that the best combination is the one with lower errors in RMSE, RE_μ, RE_б and percentage change in maximum. Hence, we may say that lower the errors, better is the combination and better is the model calibrated.

4 Conclusion

In this study, a novel approach is being tried to determine the best fit dataset for calibrating a statistical downscaling model for short-duration dataset in data scarce regions. Five IMD stations, situated in the northeastern region of India are considered for the study. To find out the effectiveness and related challenges of calibration–validation datasets, six different combinations were tried. Statistical parameters are evaluated for all the six combinations at each station to determine the correlation between calibration and validation sets. The parameters considered for the analysis are Pearson correlation, coefficient of determination (R^2), root mean square error (RMSE), relative error in mean (RE_μ), relative error in standard deviation (RE_б) and percentage change in the maximum value of the series. Based on the results, an attempt is made to find out the best combination of data for calibration as well as validation that will precisely predict the future pattern of rainfall in data scares areas. The results suggest that the best combination is the one with higher R^2 value and lower errors in RMSE, RE_μ, RE_б and percentage change in maximum. Hence, we may say that lower the errors, better is the combination and better is the model calibrated.

Conflict of Interest. The authors declare that there is no conflict of interest.

References

1. Deka SK, Sarma AK (2011) Impact of climate change on precipitation characteristics of Brahmaputra basin. A study report, IIT Guwahati
2. Gosain AK, Rao S, Basuray D (2006) Climate change impact assessment on hydrology of Indian river basins. Curr Sci 90(3):346–353
3. Lin HY, Hu JM, Chen TY et al (2018) A dynamic downscaling approach to generate scale-free regional climate data in Taiwan. Taiwania 63:25
4. Mehrotra D, Mehrotra R (1995) Climate change and hydrology with emphasis on the Indian subcontinent. Hydrol Sci J 40:231–242
5. Mujumdar PP (2008) Implications of climate change for sustainable water resources management in India. Phys Chem Earth 33:354–358
6. Sarma AK, Hazarika J (2014) GCM based fuzzy clustering to identify homogeneous climatic regions of northeast India. WASET Int J Environ Ecol Geol Geophys Eng 8(12):732–739
7. Sarma AK, Sarma PK, Vinnarasi R (2012) Climatic data collection from tea garden and other sources of northeast India for climate change study. Report submitted to Climate Change Directorate of MoWR, Govt. of India

8. Wagner PD, Reichenau TG, Kumar S, Schneider K (2015) Development of a new downscaling method for hydrologic assessment of climate change impacts in data scarce regions and its application in the Western Ghats, India. Reg Environ Change 15:435–447
9. Wilby RL, Hassan H, Hanaki K (1998) Statistical downscaling of hydrometeorological variables using general circulation model output. J. Hydrol 205:1–19
10. Wilby RL, Wigley TML, Conway D, Jones PD, Hewitson BC, Main J, Wilks DS (1998b) Statistical downscaling of general circulation model output: a comparison of methods. Water Resour Res 34(11):2995–3008
11. Zhang XC (2005) Spatial downscaling of global climate model output for site-specific assessment of crop production and soil erosion. Agric For Meteorol 135:215–229
12. Zuo J, Xu J, Chen Y, Wang C (2019) Downscaling precipitation in the data-scarce inland river basin of Northwest China based on earth system data products. Atmosphere 10(10):613

Check for updates

Chapter 6
Investigation of Climate Extremes: A Study in Dudhnoi River Basin, India

Rahul Singh Waikhom[1](✉), Nitesh Patidar[2], and Annu Taggu[3]

[1] Center for Flood Management Studies, NIH, Guwahati 781006, India
wrsingh.nihr@gov.in

[2] Ground Water Hydrology Division, NIH, Roorkee 247667, India

[3] Department of Agricultural Engineering, NERIST, Nirjuli 791109, India

Abstract. Climate change is a burning research topic in present century. A case study has been conducted to investigate the climate change in Dudhnoi river basin—a sub-watershed of the mighty Brahmaputra River through climate change indicators using daily gridded temperature and precipitation data. Determination of trends in the climate extremes instead of trends in climatic means is one of the main objectives in this study, since extreme events as a result of climate change are more intensive and cause catastrophes. Warming in both day and night temperatures along with the increase in consecutive dry days is the highlight of this study. Also, reduction in annual precipitation and number of consecutive wet days while increasing the occurrence of intense and heavy precipitation amount in short duration of time are the other highlights of the study. Hence, it draws the attention of policymakers to endorse appropriate watershed management plans with robust techniques to tackle climate extremes in the future.

Keywords: Climate change · Climate change indicators · Climate extremes · Extreme events

1 Introduction

Of all the natural disasters, disasters caused by extreme meteorological events in the world contribute approximately 85% [8]. The natural disasters in the world are due to extreme events and not just a variation of climatic means [10] and have drawn a lot of attention to study the climate extremes. The report of the Intergovernmental Panel on Climate Change [3] stated that since 1950 global average maximum and minimum temperatures over land surface have increased by an excess of 0.1 °C per decade. In the context of India, Pant and Kumar [9] found an increasing trend in mean annual temperature at the rate of 0.57 °C per 100 years and [6] found that the mean maximum temperature has raised by 0.6 °C while the mean minimum temperature has decreased by 0.1 °C.

R. K. Bhattacharjya et al. (eds.), *Sustainable Water Resources Management*,
Advances in Sustainability Science and Technology,
https://doi.org/10.1007/978-981-16-7535-5_6

Various climate change indicators have been developed in the past and being used for monitoring extreme climatic events such as floods and droughts. The changing trends of climate extremes required greater importance than trends of mean climate as extreme events as a result of climate change are more intensive and cause catastrophes [4]. A lot of efforts were being made by researchers to determine climate extreme indices based on the probability of occurrence or threshold exceedance [7]. Expert Team on Climate Change Detection and Indices (ETCCDI) found consensus on a two-pronged approach to promote further work on the monitoring and analysis of daily climate records to identify trends in extreme climate events [18]. The first prong was to document the exact formulation of an internationally agreed suite of climate extreme indices from daily precipitation and temperature data. The second prong was to extend the approach globally especially in less developed countries. The expert team identified 27 core indices based on daily temperature and daily precipitation amount and are actively used in various studies [1, 4, 5, 11, 12, 16, 17]. Tirkey et al. [16] assessed the climate extremes over Jharkhand, India, and found that the average consecutive dry days have increased during the study period. In order to understand drought characteristics and climate extremes, a comprehensive study is required in the northeastern India. Considering the above-mentioned facts, a comprehensive analysis is performed in the present study to investigate climate extremes using extreme climate indices.

2 Materials and Methods

Study Area and Data

Dudhnoi River—a tributary to the River Brahmaputra—flows through Assam and Meghalaya states of India. The study area has a geographical extent of 476 km^2, out of which approximately 20% of the area lies in flood plain region and the rest lies in the hilly terrain. Precipitation, maximum and minimum temperature for the period of 1981 to 2013 were acquired from India Meteorological Department (IMD). The obtained data sets are gridded 0.25° by 0.25° for precipitation and 1° by 1° for maximum and minimum temperatures on daily timescale. Dudhnoi river basin consists of five administrative divisions, namely Songsak, Rongjeng, Kharkutta, Mendipathar and Dudhnoi as illustrated in Fig. 1. The annual average and standard deviation of precipitation and temperature for each of the divisions are illustrated in Table 1. The annual average precipitation of all the divisions in the study area was well above the national average rainfall of 119 cm. Among the divisions, there are some variations in the precipitation but the temperature does not vary much.

Methodology

The aim is to carry out a reconnaissance study of the area with regard to climate extremes. Droughts and floods are two paradoxes of a natural hazard which are equally important. In this study, trends of climate extremes in the basin were investigated using extreme climate indices. A total of 29 grids for precipitation and 15 grids for temperature were required to cover the study area. A self-explained flow diagram describing the methodology in this study is presented in Fig. 2. Expert Team on Climate Change Detection and

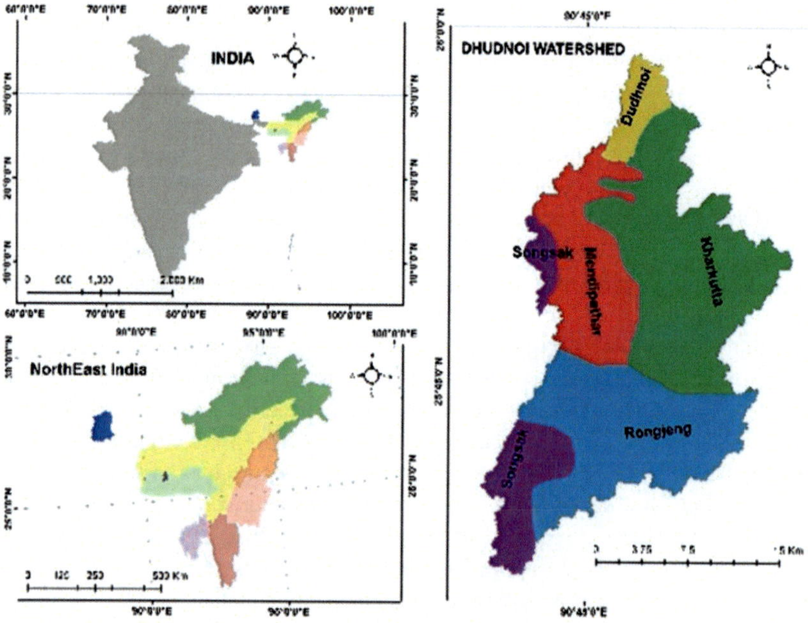

Fig. 1 Dudhnoi watershed and its five divisions

Table 1 Statistics of annual precipitation, maximum temperature (Tmax) and minimum temperature (Tmin) during 1981–2013

Division	Precipitation (mm)		T_{max} (°C)		T_{min} (°C)	
	Mean	STD	Mean	STD	Mean	STD
Songsak	2926.30	850.06	27.99	3.43	18.18	5.36
Rongjeng	2771.07	775.45	27.96	3.44	18.15	5.35
Kharkutta	2583.73	595.84	27.83	3.46	18.04	5.34
Mendipathar	2527.73	590.67	27.85	3.46	18.05	5.34
Dudhnoi	2380.37	420.98	27.71	3.48	17.94	5.33

Indices (ETCCDI) recommended 27 core climate extreme indices which can be calculated using temperature and precipitation data. Out of 27 core indices, 6 indices related to temperature and 7 indices related to precipitation were analyzed using RClimDex-1.9-3 software downloaded from https://github.com/ECCC-CDAS/RClimDex/releases/tag/1.9-3. All the indices under consideration are illustrated in Table 2. Bothale and Katpatal [1] mentioned that frequency of cold days, warm days, cold nights and warm nights is the result of extreme conditions which made them good indicators of climate change. Quality control of the data is very important for any analysis, therefore, such analysis was performed in RClimDex to flag those data which are either missing or outliers. Thus, maximum and minimum temperature data greater than 70 °C were removed from the time series before calculating extreme climate indices. An excel template MAKESENS 2.0 beta version developed by Salmi et al. [14] was used to detect trends and determine the magnitude of the trends.

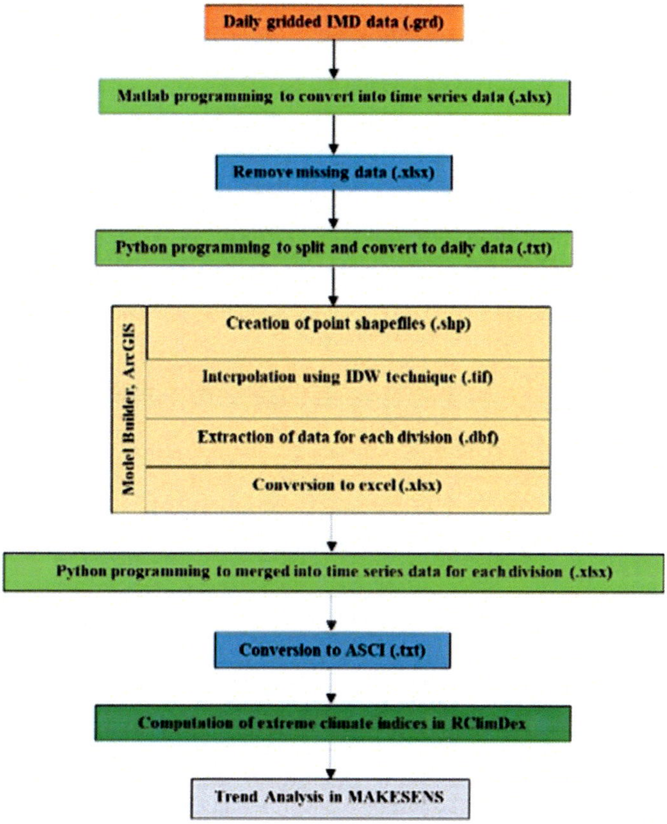

Fig. 2 Methodology flow diagram of the study

3 Results and Discussion

Trends in Temperature Extremes During 1981 to 2013

The trend analysis on the number of summer days in a year for which daily maximum temperature is greater than 20 °C (SU25) showed positive trends in all the divisions (Table 3). Kharkutta division shows positive significant trend which increases at the rate of 0.55 days per year. On the other hand, annual counts when daily minimum temperature is greater than 20 °C (TR20), which is a criterion for tropical nights, have found to follow a rapid increasing trend at the rate of more than ~0.7 days per year in all the divisions. This result is also in agreement with the significant cooling trends of cool nights (TN10p)

Table 2 Summary of extreme climate indices in the study

Index	Descriptive name	Definitions	Units
SU25	Summer days	Annual count when TX (daily maximum) >25 °C	Days
TR20	Tropical nights	Annual count when TN (daily minimum) >20 °C	Days
TN10p	Cool nights	Annual percentage of days when TN < 10th percentile	Days
TX10p	Cool days	Annual percentage of days when TX < 10th percentile	Days
TN90p	Warm nights	Annual percentage of days when TN > 90th percentile	Days
TX90p	Warm days	Annual percentage of days when TX > 90th percentile	Days
Rx5day	Max 5-day precipitation amount	Annual maximum consecutive 5-day precipitation	mm
SDII	Simple daily intensity index	Annual total precipitation divided by the number of wet days (defined as PRCP ≥ 1.0 mm) in the year	mm/day
R20	Number of very heavy precipitation days	Annual count of days when PRCP ≥ 20 mm	Days
CDD	Consecutive dry days	Maximum number of consecutive days with daily rainfall (RR) < 1 mm	Days
CWD	Consecutive wet days	Maximum number of consecutive days with RR ≥ 1 mm	Days
R95p	Very wet days	Annual total PRCP when RR > 95th percentile	mm
PRCPTOT	Annual total wet-day precipitation	Annual total PRCP in wet days (RR ≥ 1 mm)	mm

and significant warming trends of warm nights (TN90p). Annual percentage of cool days (TX10p) is decreasing, though not significant, whereas annual percentage of warm days is significantly increasing in the study area. Hence, there is a simultaneous warming of both days and nights which are statistically significant at 5% level. As observed from Sen's slope values (Q) of TR20, TN90p and TX90p in Table 3, the warming of the area during nights occurred at a higher rate than warming during days. These results are in agreement with the results of other previous studies, such as by Keggenhoff et al. [4] in Georgia, Bothale and Katpatal [1] in Godavari basin (India) and Toure et al. [17] in Mali.

This study has shown that the number of cold nights and days has decreased and the number of warm nights and days has increased. Such warming trends in both days and nights will increase domestic energy consumption due to rise in energy utilization for

Table 3 Mann–Kendall's test Z and Sen's slope Q of extreme indices (significant at 0.05 level are bold)

Indices	Dudhnoi		Songsak		Rongjeng		Kharkutta		Mendipathar	
	Z	Q	Z	Q	Z	Q	Z	Q	Z	Q
SU25	1.92	0.53	1.70	0.48	1.67	0.50	**1.98**	0.55	1.84	0.55
TR20	**2.93**	0.71	**3.19**	0.82	**3.04**	0.75	**2.93**	0.69	**3.01**	0.75
TN10p	**−2.77**	−0.42	**−2.87**	−0.41	**−2.87**	−0.41	**−2.85**	−0.42	**−2.73**	−0.41
TN90p	**3.58**	0.45	**3.77**	0.43	**3.73**	0.43	**3.67**	0.43	**3.64**	0.43
TX10p	−0.33	−0.02	−0.48	−0.04	−0.48	−0.03	−0.43	−0.04	−0.53	−0.03
TX90p	**3.05**	0.34	**2.99**	0.31	**3.11**	0.32	**3.16**	0.34	**3.11**	0.33
SDII	**2.45**	0.12	**2.01**	0.13	**2.08**	0.12	**2.36**	0.12	**2.42**	0.12
R95p	**2.08**	14.58	1.52	11.93	1.74	13.68	**2.08**	14.19	**2.11**	14.50
R20	1.18	0.25	0.57	0.19	0.43	0.12	1.08	0.23	1.01	0.20
PRCPTOT	−0.14	-1.71	−0.29	−4.18	−0.39	-7.07	−0.29	−2.27	−0.11	−1.45
CWD	**−3.41**	-1.39	**−4.31**	-1.85	**−4.34**	-1.86	**−4.17**	−1.89	**−3.69**	**−1.63**
CDD	1.92	1.02	1.91	0.89	1.91	0.87	**2.51**	1.17	**2.12**	1.00
Rx5day	1.53	1.79	1.35	2.94	1.47	3.21	1.63	2.62	1.78	2.67

cooling, increase water demands in agriculture due to increase in evaporation rate and also increase in the mortality rate [7, 15, 17]. Rao et al. [11] found that decadal trend in global temperatures has been increasing since the 1970s which is also consistent with our findings. By averaging Sen's slope Q values over the divisions, the rate of change of SU25, TR20, TN10p, TN90p, TX10p and TX90p is 0.52 days/year, 0.74 days/year, –0.41% days/year, 0.43% days/year, –0.03% days/year and 0.33% days/year, respectively.

Trends in Precipitation Extremes During 1981 to 2013

There has been significant increasing trend in simple daily intensity index (SDII) in the area approximately at the rate of 0.12 mm/day per year (Table 3). The increasing trend may be attributed to either increase in annual total precipitation represented by PRCPTOT or decrease in number of total wet days in a year. However, the trend analysis of PRCPTOT shows the annual total precipitation in the area to follow a decreasing trend. Therefore, increasing trend in SDII is due to decrease in the wet days. Also, the decreasing trend of consecutive wet days (CWDs) and increasing trend of consecutive dry days (CDDs) partly support the positive trend of SDII. Shrinkage of monsoon season, i.e., late onset and early withdrawal of southwest monsoon as well as the increase in dry days, could possibly be the causes of falling trends of PRCPTOT [2, 16]. The trend of very wet days (R95p) shows statistically significant positive trend in Dudhnoi, Kharkutta and Mendipathar divisions. This indicates that although precipitation amount and number of wet days have decreased, the precipitation is becoming more and more intense. This statement is also supported by the positive trends of annual counts of days when precipitation is greater than 20 mm (R20) and annual maximum consecutive 5-day precipitation amount (Rx5day) as shown in Table 3. Despite insignificant trends in R20 and Rx5day, it can unarguably state that precipitation of the nature of short duration high intensity is occurring very frequently in the area. This tendency of amplifying response of precipitation extremes such as higher intensity and large amount of precipitation in very short period of time is consistent with the increase of extreme precipitation events globally [4, 13]. By averaging Sen's slope Q values over the divisions, the rate of change of SDII, R95p, R20, PRCPTOT, CWD, CDD and Rx5day is, respectively, 0.12 mm/day/year, 13.78 mm/year, 0.20 days/year, -3.34 mm/year, –1.72 days/year, 0.99 days/year and 2.65 mm/year. The three precipitation extreme indices CWD, CDD and Rx5day are plotted as shown in Fig. 3.

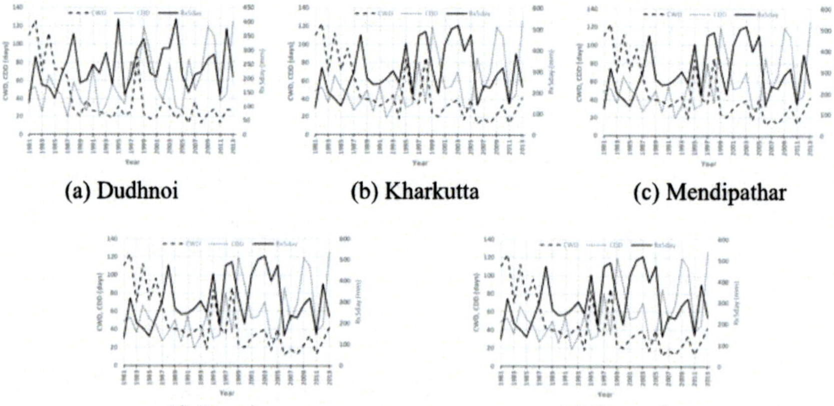

(a) Dudhnoi (b) Kharkutta (c) Mendipathar

(d) Rongjeng (e) Songsak

Fig. 3 Consecutive wet days (CWDs), consecutive dry days (CDDs) and maximum 5-day precipitation amount (Rx5day) for **a** Dudhnoi, **b** Kharkutta, **c** Mendipathar, **d** Rongjeng and **e** Songsak divisions

In the figure, CWD is denoted by dashed line, CDD by dotted lines and Rx5day by solid lines. The consecutive wet days are found to be higher (around 100 days) in the initial years, but in the later end of the study period, CWD becomes lower almost below 40 days. In case of consecutive dry days, the trends are found exactly the opposite of CWD. This rapid rise and fall of CCD and CWD, respectively, are warning signal of future water scarcity, and it can bring an alarming situation of socioeconomic and hydrological condition. This study points out the necessity to develop proper water management plans, for example, storing excess water during events of heavy rainfall to utilize later during events of droughts. The maximum 5-day precipitation in the study area ranges from 150 to 500 mm for all station except Dudhnoi that ranges from 115 to 425 mm. Despite being insignificant trends in maximum 5-day rainfall in all stations, the decrease in CWD and increase in CDD indicate water stress conditions, i.e., drought. This is due to the fact that the reduction in annual precipitation and number of CWD occurred along with increase in occurrence of higher intensity and lower duration precipitation.

4 Conclusion

The present study highlights the variability of climate extreme indicators over the five divisions of Dudhnoi river basin. RClimDex software was used to determine the climate extremes of both precipitation and temperature. It is worth to mention that the variability of climate extremes among the five divisions is negligible. The number of tropical nights (TR20) and percentages of warm nights (TN90p) and warm days (TX90p) are all significantly (at 5% level) increased between 1981 and 2013, while the percentage of cool nights (TN10p) is significantly decreased during the same period. Also, a number of summer days have increased and percentage of cool days (TX10p) has decreased between the study periods although not significant at 5% level. An increasing trend at 5% significant level is found for SDII (all divisions), R95p (Dudhnoi, Kharkutta and Mendipathar) and CDD (only in Kharkutta). On the other hand, falling trend is observed for CDD at same significant level in all the divisions. The number of very heavy precipitation (R20) and maximum 5-day precipitation amount (Rx5day) have increased between the study periods although they are not significant. The insignificant decreasing trends in annual precipitation (PRCPTOT) are also observed which is accompanied by the increasing trends of consecutive dry days (CDDs). Despite the decreasing trends in PRCPTOT, increasing trends in R95p, R20, SDII and Rx5day are observed signifying the climate of the area moving toward more extremes. The analysis of dry days during southwest monsoon concluded that Dudhnoi division has been found to be the most affected division followed by Rongjeng, Mendipathar, Kharkutta and Songsak. This case study exposes the need to conserve water through development of water storage structures to be utilized during the lean periods. It also brings the need to pay more attention to various climate extremes which are probable in the future through proper monitoring, prevention and preparedness. The closure of this study is that we can expect extreme climate events such as warming, intense rainfall and cloudburst to occur more frequently in the future accompanied by long spells of dry period adding pressure on water resources and therefore should be prepared for such unpleasant climate extremes.

Conflict of Interest. None.

References

1. Bothale RV, Katpatal YB (2015) Trends and anomalies in extreme climate indices and influence of El nino and La Ni ˜ na over Pranhita Catchment in Godavari Basin, India. J Hydrol Eng 21(2)
2. Guhathakurta P, Rajeevan M (2008) Trends in the rainfall pattern over India. Int J Climatol 28:1453–1469
3. IPCC (Intergovernmental Panel on Climate Change) (2013) Climate change 2013—the physical science basis, working group I contribution to the IPCC fifth assessment report (WGI AR5) of the intergovernmental panel on climate change. Cambridge University Press, pp 422–808
4. Keggenhoff I, Elizbarashvili M, Amiri-Farahani A, King L (2014) Trends in daily temperature and precipitation extremes over Georgia, 1971–2010. Weather Clim Extremes 4:75–85
5. Kim BS, Yoon YH, Lee HD (2011) Analysis of changes in extreme weather events using extreme indices. Environ Eng Res 16(3):175–183

6. Kumar KR, Kumar KK, Pant GB (1994) Diurnal asymmetry of surface temperature trends over India. Geophys Res Lett 21(8):677–680
7. Ly M, Traore SB, Alhassane A, Sarr B (2013) Evolution ofsome observed climate extremes in the West African Sahel. Weather Clim Extreme 1:19–25
8. Obasi GOP (1994) WMO's role in the international decade for natural disaster reduction. Bull Am Meteorol Soc 75(9):1655–1661
9. Pant GB, Kumar KR (1997) Climates of South Asia. Wiley, Chichester, UK
10. Plummer N, Salinger MJ, Nicholls N, Suppiah R, Hennessy KJ, Leighton RM, Trewin B, Page CM, Lough JM (1999) Changes in climate extremes over the Australian region and New Zealand during the twentieth century. Clim Change 42(1):183–202
11. Rao KK, Patwardhan SK, Kulkarni A, Kamala K, Sabade SS, Kumar KK (2014) Projected changes in mean and extreme precipitation indices over India using PRECIS. Global Planet Change 113:77–90
12. Revadekar JV, Kothawale DR, Patwardhan SK, Pant GB, Kumar KR (2012) About the observed and future changes in temperature extremes over India. Nat Hazards 60(3):1133–1155
13. Roy SS, Balling JR (2004) Trends in extreme daily precipitation indices in India. Int J Climatol 24(4):457–466
14. Salmi T, Maatta A, Anttila P, Ruoho-Airola T, Amnell T (2002) Detecting trends of annual values of atmospheric pollutants by the Mann-Kendall Test and sen's slope estimates-the excel template application Makesens. Finnish Meteorological Institute Publications on Air Quality No. 31, Helsinki, Finland
15. Thornton PK, van de Steeg J, Notenbaert A, Herrero M (2009) The impacts of climate change on livestock and livestock systems in developing countries: a review of what we know and what we need to know. Agric Syst 101:113–127
16. Tirkey AS, Ghosh M, Pandey AC, Shekhar S (2018) Assessment of climate extremes and its long term spatial variability over the Jharkhand state of India. Egypt J Remote Sens Space Sci 21(1):49–63
17. Toure HA, Kalifa T, Kyei-Baffour N (2017) Assessment of changing trends of daily precipitation and temperature extremes in Bamako and Segou in Mali from 1961–2014. Weather Climate Extremes 18:8–16
18. Zhang X, Alexander L, Hegerl GC, Jones P, Tank AK, Peterson TC, Trewin B, Zwiers FW (2011) Indices for monitoring changes in extremes based on daily temperature and precipitation data. WIREs Clim Change 2:851–870

Modelling

Chapter 7
Assessment of Land Use Change Impact on Sediment Yield Using SWAT and Partial Least Squares Regression Model

Alemayehu A. Shawul[1(✉)] and Chakma Sumedha[2]

[1] Assistant Professor, Natural Resource Management Department, Madda Walabu University, Bale Robe, Ethiopia

[2] Civil Engineering Department, Indian Institute of Technology Delhi, Hauz Khas, New Delhi 110016, India

Sumedha.Chakma@civil.iitd.ac.in

Abstract. A holistic understanding of the impact of land use/land cover (LC) changes on the sedimentations is important for efficient decisions on sustainable watershed management practices. Integrated application of hydrologic modeling and partial least squares regression (PLSR) model was applied to quantify the impact of anthropogenic activities on sediment yield in the Upper Awash basin. The soil and water assessment tool (SWAT) model performed well for simulation of monthly sediment yield at four main gauged subbasins. More than 55% of the total annual sediment yield occurs in the main rainfall season (JJA) peaking in July and August. Larger areas of the basin are experiencing beyond the tolerable limit of soil loss, with more than 39% area of the basin generates >20 t/ha/year. The effect of land cover change on sediment yield is highly pronounced at the subbasins scale than the whole basin scale. Moreover, the watershed area which is under higher soil loss severity increases with the change in land cover. The PLSR correlation matrix, variable importance for the projection (VIP), and the standardized PLSR coefficient exhibited that sediment yield is highly attributed to the change in cropland and pasture land. Pasture and cropland demonstrated the higher VIP values of more than 1.0. Thus, the variables were considered more critical for the prediction of sediment yield. While shrubland and water body were relatively less important factors affecting sediment yield, which resulted in a lower VIP value of less than 0.5. It also exhibited that encroachment of cropland and urban area at the expense of forest and grassland significantly triggered sediment yield. Hence, future land use plans should consider appropriate vegetative measures and reforestation of the upland areas to reduce sedimentation. PLSR model is a useful statistical tool for assessment of the relative impacts multiple land management practices on sediment yield.

Keywords: Sediment yield · Land cover change · SWAT model · PLSR model · Upper Awash basin

R. K. Bhattacharjya et al. (eds.), *Sustainable Water Resources Management*, Advances in Sustainability Science and Technology,
https://doi.org/10.1007/978-981-16-7535-5_7

1 Introduction

The high rate of population growth, unplanned land use system, relying on persistent agriculture and natural resource leads to progressive deforestation, accelerated soil erosion with the corresponding decrease in soil fertility [1, 2]. The high rate of soil erosion results in loss of soil productivity, environmental pollution, and siltation of water bodies and crop fields. Siltation is a severe problem in Ethiopia and is currently causing a shortage in domestic water and hydroelectric power supply, further creating hindrance mainly on water resources related development programs. Farmers in the highland areas of Ethiopia cultivate hilly lands due to lack of enough arable lands causing topsoil to severely wash away during the rainy season [3, 4]. Hence, farming of unsuitable land for agriculture, declining crop production leading to reduced production and create further pressure on natural resources. The continued agricultural soil erosion and soil loss become a critical problem for global agricultural production under conventional farming practices,thus, the global society should adopt methods that sustain soil fertility and crop production [5]. Many natural and anthropogenic factors affect soil erosion, including climatic characteristics, soil properties, land use, and topography [6]. Soil properties and topography can be considered unchanged in the short term, whereas, the land cover (LC) dynamics and climatic features are the dominant variables causing soil erosion [7]. Soil erosion and subsequent sediment yield are highly related to the spatial configuration of the LC types in the watershed [8].

Improved tillage practices and adoption of conservation tillage, such as no-tillage, could strongly help to reduced sediment yields in the watersheds [9]. The high soil erodibility factors and the persistent cultivation of cereal crops mainly *teff* lead to severe siltation of reservoirs, reducing its original volume and producing several sediment deposits in the Upper Awash basin [3, 4, 10]. Assessment of the impacts of LC change on runoff variability and sedimentation is an exacting research problem. The impact of best management practices and alternative land use scenarios on the water quantity and quality can be assessed effectively using integrated application of soil and water assessment tool (SWAT) and partial least squares regression (PLSR) model. Therefore, the objective of this study is (i) to determine the influences of land use or/and land cover (LC) change on the sediment yield at main gauged subbasins and (ii) to identify the relative contribution of change in the LC classes on sediment yield based.

2 Materials and Methods

Sediment Yield Data Sources

Daily measured discharge and suspended sediment samples were obtained from the hydrology department of the Ministry of Water, Irrigation and Energy of Ethiopia (MoWIE) for the main gauging stations. A sediment rating curve is mainly applied to obtain the value of sediment concentration for a given river discharge data based on sediment sample data. The sediment rating curve was fitted to estimate continuous daily sediment yield for four main stations in the Upper Awash basin, namely Hombole, Melka Kuntre, Mojo and Akaki stations. The monthly total sediment yield was determined based on the daily sediment series. The SWAT model input data mainly the climate data, soil data, LULC data and topographic data were used for simulation of sediment yield. The daily meteorological data were obtained from National Meteorological Agency (NMSA) (Fig. 1).

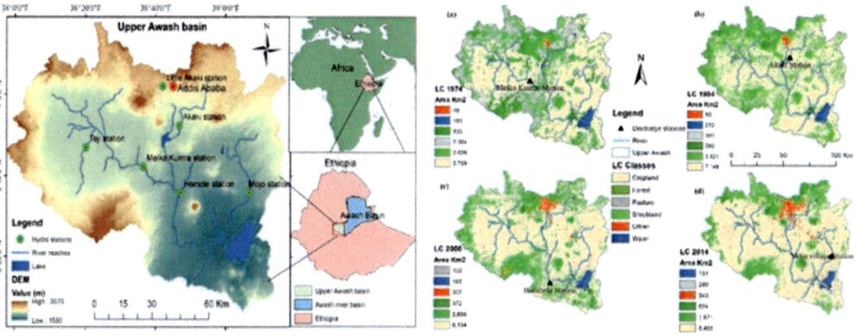

Fig. 1 Location map of Upper Awash basin in the central part of Ethiopia; and LC 1974, LC 1984, LC 2000 and LC 2014 in the Upper Awash basin

SWAT Model Setup for Sediment Simulation

Simulation of Sediment Yield

The SWAT model is semi-distributed computationally efficient and capable of simulating streamflow, sediment yield, nutrients, and pesticide transport at different spatial and temporal scales [11, 12]. The SWAT watershed model contains algorithms for simulating erosion from each hydrologic response units (HRUs) of the watershed. Erosion caused by rainfall and runoff is estimated using the Modified Universal Soil Loss Equation (MUSLE) [13] indicated in Eq. 1. MUSLE computes sediment yield based on the Universal Soil Loss Equation (USLE) soil erodibility factor, the cover and management factor, the support practice factor, the topographic factor, and a coarse fragment factor and from the surface runoff volume, the peak runoff rate, and the area of the HRUs. In MUSLE, the rainfall energy factor is replaced with a runoff factor. It improves the sediment yield prediction, eliminates the need for delivery ratios, and allows the equation to be applied to individual storm events. Sediment yield prediction is improved because runoff is a function of antecedent moisture conditions as well as rainfall energy. Delivery ratios are not needed with MUSLE because the runoff factor represents the energy used in detaching and transporting sediment [14]. The detailed SWAT model input/output documentation, theoretical backgrounds, software, and repository of literature databases can be obtained from the website https://swat.tamu.edu/.

$$\begin{aligned} \text{Sed} = {} & 11.8(Q_{\text{surf}} \times q_{\text{peak}} \times \text{area}_{\text{hru}})^{0.56} \times K_{\text{USLE}} \times C_{\text{USLE}} \\ & \times P_{\text{USLE}} \times \text{LS}_{\text{USLE}} \times \text{CRFRG} \end{aligned} \quad (1)$$

where Sed is the sediment yield (metric tons), Q_{surf} is the surface runoff volume (mm/ha), Q is the peak rate of runoff (m^3/s), area HRU is the area of the HRU (ha), K_{USLE} peak is the USLE soil erodibility factor (0.013 metric ton m^2 h/(m^3-metric ton cm)), C_{USLE} is the USLE cover and management factor, P_{USLE} is the USLE support practice factor, LS_{USLE} is the USLE topographic factor, and CFRG is the coarse fragment factor.

The sediment calibration and validation were performed at four main subbasins in the Upper Awash basin, namely: Hombole, Melka Kuntre, Mojo, and Akaki subbasins. Calibration of the sediment yield was performed by iteratively changing the sensitive parameters within the allowable minimum and maximum bounds. The goodness of fit between observed and simulated sediment yield was determined using correlation coefficient (R^2), Nash–Sutcliffe efficiency (NSE), percent bias (PBIAS) and the RMSE-observations standard deviation ratio (RSR) of measured and simulated sediment. The observed sediment data period which has relatively consistent hydrologic and sediment data series were selected for calibration and validation of the model. The calibrated and validated model was run separately to evaluate the impacts of land cover (LC) change on the sediment yield (historical LC 1974, 1984, 2000, and 2014). Temporal and spatial LC change detection in the Upper Awash basin between the year 1974 and 2014 has been documented in Shawul and Chakma [15, 16]. Long-term sediment load and sediment concentration simulation from 1980 to 2014 was undertaken on a monthly and annual time scales for different LC inputs. Moreover, analysis of the impacts of LC change on the sediment fluxes was performed to evaluate the significance of LC change impact at different spatial and temporal scales.

Application of PLSR for Analysis of LC Change Impact on Sediment Yield

Partial least squares regression (PLSR) is a robust multivariate regression method that can be used to perform a wide range of analyses when the predictors exhibit collinearity [8, 17]. PLSR relates two data matrices x and y by a linear multivariate model,each model parameter is estimated as the slope of a simple bivariate regression (least squares) between a matrix column or row as the y-variable, and another parameter vector as the x-variable [17]. Detailed statistical background of PLSR modeling can be found in the literature such as [17, 18]. PLSR is an extension of a multiple linear regression model. In the simplest form, a linear model specifies the relationship between a dependent variable y and a set of predictor variables x as shown in Eq. 2.

$$y=b_0+b_1x_1+b_2x_2+\cdots+b_px_p \tag{2}$$

where b_0 is the regression coefficient for the intercept, and the b_i values are the regression coefficients (for variables 1 through p) computed from the data. In the PLSR modeling, the importance of a predictor for the dependent variables is indicated by the variable importance in the projection (VIP) (Yan et al. 2013). The factors with high VIP values of more than 1 are the most relevant in explaining the dependent variable and are considered to be significantly influential predictors of the PLSR model (Shi et al. 2014; Li et al. 2017). Wold (1995) suggests VIP value more than 0.8 is most relevant and significant for explaining the dependent variable, and values less than 0.5 are considered to be less important. The VIP and standardized regression coefficients were used to explain the relative influence of each independent variable. Thus, it is possible to determine the individual LC classes (predictor variables) which strongly interact with the corresponding sediment yield (the dependent variable). The VIP values provide information on the strength of influence for each predictor, and the regression coefficients give the direction of the influence [19, 20]. The PLSR model was used to evaluate the relationship between the changes in annual sediment yield and particular LC classes. The change in the LC maps of 1974, 1984, 2000 and 2014, and mean annual sediment yield simulated based on each LC maps was used.

3 Results and Discussion

Calibration and Validation of Sediment Yield

Sensitivity analysis, model calibration, and validation for sediment yield were performed using the Sequential Uncertainty Fitting (SUFI2) algorithm in the SWAT-CUP program [21]. Sensitive sediment parameters were used for calibration and validation of the model for the sediment yield at the main gauged subbasins in the Upper Awash basin. The channel cover factor (CH_COV1), channel erodibility factor (CH_ERODMO), USLE equation soil erodibility (K) factor (USEL-K), USLE cover factor (USLE-C), USLE support practice factor (USLE-P), average slope length (SLSUBBSN), slope length for lateral subsurface flow (SLSOIL), linear factor for channel sediment routing (SPCON), and exponential factor for channel sediment routing (SPEXP) are among the most sensitive sediment parameters.

The result from statistical model performance evaluation measures met the requirement of NS > 0.6, $R^2 > 0.6$, which was set by model criteria such as [22, 23]. The value of RSR efficiency measure was maintained at less than 0.7 for all stations both at calibration and validation periods, which indicates good to satisfactory model performances, as shown in Table 1. The percent bias (PBIAS) between the simulated and observed sediment yield also indicated good performance during the calibration periods at all stations maintained value less than 10% except at Mojo station, which was found to be 14.5%. The statistical measures of simulated versus observed monthly sediment yield show better performance at Hombole, Melka Kuntre, Mojo, and Akaki subbasins. The model performance for sediment yield simulation could be improved with more refined sediment sample measurements and model input data. Overall, the monthly sediment simulation analysis indicated that the SWAT model performed well to the best of its ability. Thus, the SWAT model can be considered as a capable tool for simulating the sediment yield for different plausible land management scenarios.

Impact of Land Cover Change on the Sediment Yield

Planning for soil and water conservation requires knowledge of the relationships between factors that are responsible for soil loss and factors that help reduce soil losses [3]. The LC change between 1974 and 1984 period revealed a higher increase in cropland (18.1%) and massive loss in vegetation cover, mainly pasture (–14.0%), as shown in Table 2. As a result, sediment yield tremendously increased by 13.2% from 104.5 to 118.3 t/ha. At all the gauged subbasins, maximum sediment yield was obtained in the JJA wet season

Table 1 Statistical model performance measures for sediment yield simulation at Hombole, Melka Kuntre, Mojo, and Akaki subbasins for calibration (Cal) and validation (Val) periods

Subbasins	Hombole		Melka Kuntre		Mojo		Akaki	
Statistical measures	Cal	Val	Cal	Val	Cal	Val	Cal	Val
R^2	0.72	0.73	0.68	0.67	0.74	0.64	0.65	0.60
NS	0.70	0.72	0.65	0.63	0.68	0.63	0.64	0.60
PBIAS	–6.16	–25.2	–3.68	8.15	–14.2	1.36	6.55	29.0
RSR	0.55	0.53	0.59	0.61	0.57	0.61	0.6	0.70

with the peak value in July and August months that comply the hydrologic patterns of the basin, as shown in Fig. 2a–d. The maximum difference in the mean monthly sediment yield was obtained at Teji subbasin 49.1% increase between LC 1974 and LC 2014. Subbasins that have experienced more significant alteration of forest and shrubland to cropland resulted in a higher increment of sediment yields such as Teji, Hombole, and Melka Kuntre subbasins. The impact of LC change on sediment yield is highly pronounced at smaller subbasin level than the whole basin. Higher seasonal change in sediment yield occurred in MAM at Hombole and Melka Kuntre subbasins, whereas in Mojo and Teji subbasins, the maximum increase of 13.0 and 37.1%

occurred in JJA season, respectively. Moreover, the seasonal variations of sediment yield exhibited that more than 55% of mean annual sediment yield occurs in JJA season which is the main rainfall season in the Upper Awash basin. In the Hombole, Melka Kuntre, Teji, and Mojo subbasins, higher increase in the sediment yield was observed due to a significant amount of shrubland, forest, and pasture land which were converted to cropland between the year 1974 and 2014. The higher spatial variation of LC classes can affect the sediment yield and runoff process. Sediment yield is highly sensitive to the dynamics of LC and various soil and water management practices in the watershed [24]. The subbasins level change in the area of land cover class in the four main gauged subbasins Hombole, Melka Kuntre, Teji, and Mojo demonstrated higher variations between LC years 1974 and 2014 [16]. The subbasins which experienced a higher loss of forest, shrubland, and pasture to cropland received significant increments in the sediment yield.

The temporal and spatial scales of assessment are important factors for understanding how land use/land cover (LC) change will impact hydrologic ecosystem services [25]. Subbasins level LC change impact on sediment yield highlights the importance of examining localized impacts at appropriate scales as the effect is highly pronounced at the subbasins scale than the overall basin scale. Subbasins and river reach level outputs were used for identifying the spatial variations of sediment yield and soil erosion for 77 subbasins of the Upper Awash basin. Most of the subbasins which are located in the middle and upstream area found to carry higher values. However, the number of subbasins generated annual sediment yield more than 30t/ha/year was higher for the recent LC 2014 as compared to LC 1974. The spatial map of subbasin level sediment yield allows identifying subbasin which is producing a higher amount of sediment is vital for effective planning of soil conservation practices. Based on the estimated rate of

Table 2 Area of LC classes (km^2), correlation matrix of predictor variables and dependent variable (sediment yield t/ha), and PLSR model quality statistics in the Upper Awash basin

Land cover (LC)	Cropland	Forest	Pasture	Shrubland	Urban	Water	SedQ
LC 1974	5759.3	833.2	2304.4	2626.1	48.6	191.0	104.5
LC 1984	7897.9	777.2	663.6	2112.9	99.3	212.3	118.3
LC 2000	8134.4	371.7	103.3	2657.8	302.2	192.6	119.2
LC 2014	8485.0	624.5	289.3	1870.6	343.1	150.9	114.8
Correlation matrix	Cropland	Forest	Pasture	Shrubland	Urban	Water	SedQ
Cropland	1.00	–0.62	–0.98	–0.59	0.80	–0.30	0.90
Forest		1.00	0.75	–0.22	–0.80	0.21	–0.64
Pasture			1.00	0.41	–0.81	0.22	–0.93
Shrubland				1.00	–0.32	0.47	–0.31
Urban					1.00	–0.69	0.56
Water						1.00	0.13
SedQ							1.00

PLSR model quality statistics

Variables	R^2	MSE	RMSE	Q^2	Comp	Explained variation in Y (%)	Cum explained variation in Y (%)	Q^2 cum
SedQ	0.80	6.72	2.59	0.56	1	76.4	76.4	0.56
					2	22.8	99.2	0.80

SedQ = sediment yield; Cum = Cumulative; Comp = PLSR Components; Q^2 cum = Q^2 cumulative

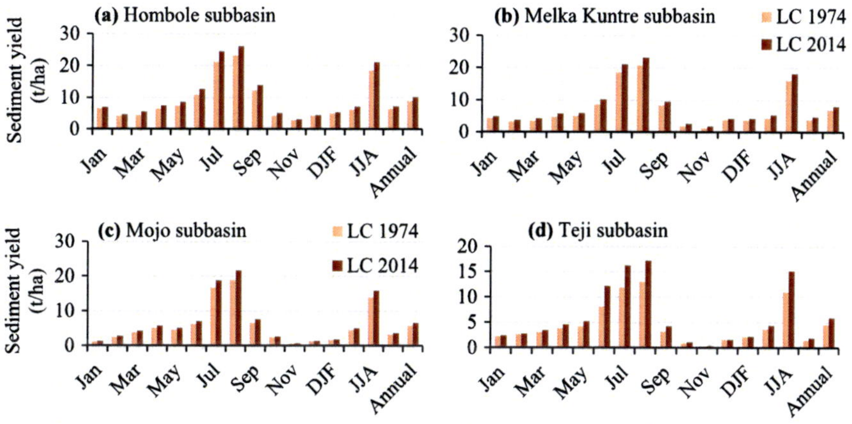

Fig. 2 Mean monthly and seasonal sediment yield for LC 1974 and LC 2014, change between LC 1974 and LC 2014 at main gauged subbasins in the Upper Awash basin

soil formation at various locations in Ethiopia, the tolerable range of soil loss levels for the different agro-ecological zone of Ethiopia is 2–18 tons/ha [2]. Nevertheless, a large area of the basin is beyond the tolerable limit of soil loss, as shown in Fig. 3. Hence, proper land management practices should be used to reduce the rate of soil erosion and sedimentation. Almost all subbasins have experienced an incremental change in the sediment yield due to LC deterioration. Mainly, the northern part of the basin encountered increasing trends of sediment yield with more 70% increase between LC 1974 and LC 1984. Subbasins that encountered an increasing trend in the area of cropland on the expenses of pasture, forest, and shrubland revealed a higher increment in sediment yield.

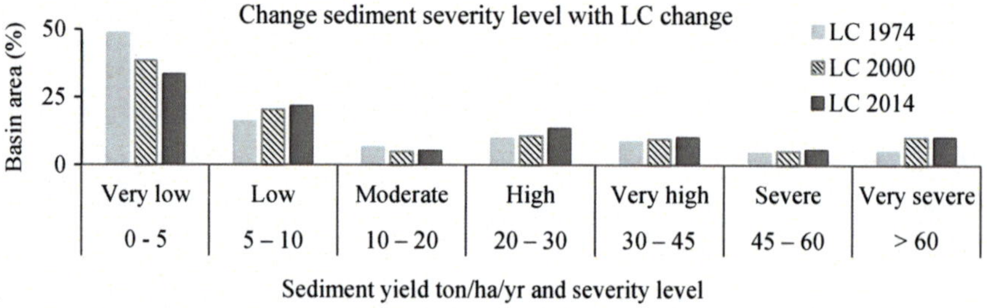

Fig. 3 Change sediment severity level for different LC inputs in the Upper Awash basin

The Relative Impact of LC Change on Sediment Yield Based on PLSR Model

The cumulative explained variation in Y and Q^2 cum that corresponds to the correlations between the explanatory (X) variable and dependent (Y) variables with the components is close to 1 with two PLSR components. This indicates the two PLSR components summarize well both the X and the Y variables as shown in Table 2. The first PLSR component explained 76.4% of the variability in the SedQ, and the cumulative explained variance for the second component was 99.2%. The correlation matrix indicated in Table 2 shows the correlations among the explanatory variables (LC classes), the dependent variables (sediment yield), and between both groups of variables. The correlation coefficient (CC) revealed that the annual sediment yield (SedQ) was highly attributed to change in the area of cropland with a CC value of 0.90. On the contrary, the SedQ was inversely related to change in the area of pasture land, forest, and shrubland resulted in CC values of –0.93, –0.64, and –0.31, respectively. However, the change in the area of forest, pasture, and shrubland has a negative correlation with surface runoff and sediment yield, which implies that an increase in vegetation covers would significantly reduce the runoff and sediment yield. In accordance with the current study, Shi et al. (2014) found that the change in the forest area is negatively related to the surface runoff amount and sediment yield. It can also be seen that change in the area of the water body has lower influence on the SedQ.

The VIP values indicated in Fig. 4a, for the first PLSR component of the six predictor variables (LC classes), shows the most important contributor to change in the annual sediment yield. The horizontal dashed lines were drawn based on the threshold value of

VIP larger than 0.8 and 1.0. Pasture and cropland demonstrated a higher VIP value of 1.44 and 1.38 more than 1.0 threshold. Thus, the variables were considered more critical for the prediction of SedQ. Moreover, if the wold's criteria of VIP >0.8 is considered, forest and urban area would be one of the effective factors for change in the sediment yield as shown in Fig. 4, whereas change in the area of shrubland and water body was less important for the prediction of SedQ. The standardized coefficient of the PLSR components which is indicated in Fig. 4b exhibits not only the magnitude but also the direction of influence of particular LC classes on sediment yield. Yan et al. (2013) have also indicated that change in the farmland and urban area extracts greater importance in the variability sediment yield. The higher loss of pasture land due to the increase of cropland between 1974 and 2014 significantly triggered soil erosion and sedimentation. Further studies on the application of PLSR model on the impact of different land uses and land management practices will help to provide a better understanding of the watershed processes that is vital for sustainable water management.

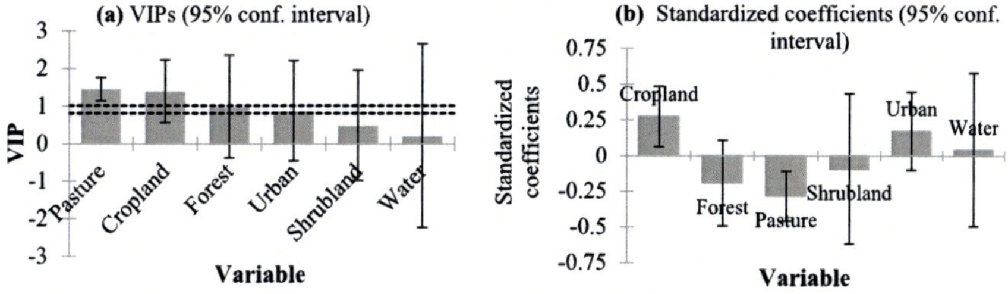

Fig. 4 VIP values of LC classes and standardized coefficients of PLSR-based analysis of land cover change impact on SedQ in the Upper Awash basin

4 Conclusion

Sustainable management of sedimentation problems largely depends on the holistic understanding of the main factors affecting soil loss and sediment source areas within the particular watershed. The relative contributions of specific land use/land cover (LC) classes to change in the sediment yield were analyzed based on hydrologic modeling and statistical tools. The SWAT model and partial least squares regression (PLSR) were applied to examine the impact of historical LC change on sediment yield quantitatively. The SWAT model performed well for monthly sediment yield simulation during the calibration and validation period at the four main subbasins. Thus, it can be considered for different scenario evaluations in the Upper Awash basin. The shift from native forest, pasture, and shrubland to cultivation land highly triggered the sediment yield. Moreover, the watershed area, which is under higher soil loss severity level, increased with the change in the LC. The impact of LC change on sediment yield is higher at smaller subbasins level than the whole basin level due to the compensating effect at the larger basins. Therefore, it is suggested to conduct related studies at small-scale subbasins to obtain reliable outcomes. The correlation matrix, the variable importance for the projection (VIP), and the standardized coefficient of PLSR exhibited that sediment yield is highly attributed to the change in cropland and pasture land. Thus, the predictor variables were considered more important for the prediction of sediment yield, whereas shrubland and water body were relatively less critical LC types affecting sediment yield. PLSR model is a useful statistical tool for quantitative assessment of the relative impacts of multiple land management practices on sediment yield, which would help to provide better decisions on watershed management.

References

1. Awulachew SB, Tenaw M, Steenhuis T, Easton Z, Ahmed A, Bashar KE (2008) Blue Nile flow, sediment and impact of watershed interventions: case of Gumera Watershed (No. 614–2016–40873)
2. Hurni H (1989) Applied soil conservation research in Ethiopia. In: Thomas DB (ed) Soil and water conservation in Kenya. University of Nairobi; Swedish International Development Authority (SIDA), Nairobi, Kenya, pp 5–21
3. Chekol DA (2006) Modeling of hydrology and soil erosion of Upper Awash River Basin (p 235). Cuvillier
4. Geleta HI (2011) Watershed sediment yield modeling for data scarce areas. PhD thesis, University of Stuttgart
5. Montgomery DR (2007) Soil erosion and agricultural sustainability. Proc Natl Acad Sci 104(33):13268–13272
6. Wischmeier WH, Smith DD (1978) Predicting rainfall erosion losses: a guide to conservation planning. Agriculture handbook 282. USDA-ARS
7. Wei W, Chen LD, Fu BJ, Huang ZL, Wu DP, Gui LD (2007) The effect of land uses and rainfall regimes on runoff and soil erosion in the semi-arid loess hilly area. China J Hydrol 335:247–258
8. Shi Z, Ai L, Li X, Huang X, Wu G, Liao W (2013) Partial least-squares regression for linking land-cover patterns to soil erosion and sediment yield in watersheds. J Hydrol 498:165–176
9. Kirsch K, Kirsch A, Arnold JG (2002) Predicting sediment and phosphorus loads in the Rock River basin using SWAT. Trans ASAE 45(6):1757
10. Halcrow SW (1989) Master plan for the development of surface water resources in the Awash basin
11. Arnold JG, Moriasi DN, Gassman PW, Abbaspour KC, White MJ, Srinivasan R, Santhi C, Harmel RD, Van Griensven A, Van Liew MW, Kannan N (2012) SWAT: Model use, calibration, and validation. Trans ASABE 55(4):1491–1508
12. Gassman PW, Reyes MR, Green CH, Arnold JG (2007) The soil and water assessment tool: historical development, applications, and future research directions. Trans ASABE 50(4):1211–1250
13. Williams JR (1975) Sediment-yield prediction with universal equation using runoff energy factor. p. 244–252. In Present and prospective technology for predicting sediment yield and sources: Proceedings of the sediment yield workshop, USDA Sedimentation Lab., Oxford, MS, November 28–30, 1972. ARS-S-40
14. Neitsch SL, Arnold JG, Kiniry JR, Williams JR (2011) Soil and water assessment tool theoretical documentation version 2009. Texas Water Resources Institute. Report No. 406. Texas A&M University, College Station, TX
15. Shawul AA, Chakma S (2019) Spatiotemporal detection of land use/land cover change in the large basin using integrated approaches of remote sensing and GIS in the Upper Awash basin, Ethiopia. Environ Earth Sci 78(5):141

16. Shawul AA, Chakma S, Melesse AM (2019) The response of water balance components to land cover change based on hydrologic modeling and partial least squares regression (PLSR) analysis in the Upper Awash Basin. J Hydrol Reg Stud 26:100640
17. Wold S, Sjöström M, Eriksson L (2001) PLS-regression: a basic tool of chemometrics. Chemometr Intell Lab 58(2):109–130
18. Abdi H (2003) Partial least square regression (PLS regression). Encycl Res Methods Soc Sci 6(4):792–795
19. Onderka M, Wrede S, Rodný M, Pfister L, Hoffmann L, Krein A (2012) Hydrogeologic and landscape controls of dissolved inorganic nitrogen (DIN) and dissolved silica (DSi) fluxes in heterogeneous catchments. J Hydrol 450:36–47
20. Woldesenbet TA, Elagib NA, Ribbe L, Heinrich J (2017) Hydrological responses to land use/cover changes in the source region of the Upper Blue Nile Basin, Ethiopia. Sci Total Environ 575:724–741
21. Abbaspour KC (2015) SWAT-CUP: SWAT calibration and uncertainty programs. Swiss Federal Institute of Aquatic Science and Technology Eawag: Duebendorf, Switzerland, pp 1–100
22. Moriasi DN, Arnold JG, Van Liew MW, Bingner RL, Harmel RD, Veith TL (2007) Model evaluation guidelines for systematic quantification of accuracy in watershed simulations. Trans ASABE 50(3):885–900
23. Santhi C, Arnold JG, Williams JR, Dugas WA, Srinivasan R, Hauck LM (2001) Validation of the swat model on a large river basin with point and nonpoint sources. J Am Water Resour Assoc 37(5):1169–1188
24. Worku T, Khare D, Tripathi SK (2017) Modeling runoff–sediment response to land use/land cover changes using integrated GIS and SWAT model in the Beressa watershed. Environ Earth Sci 76(16):550
25. Barlow JE, Burns IS, Guertin DP, Kepner WG, Goodrich D, McCarthy JM (2016) Assessing hydrologic impacts of future land cover change scenarios in the South Platte River Basin (CO, WY, & NE) and the San Pedro River Basin (US/Mexico). U.S. Environmental Protection Agency Office of Research and Development Washington, DC 20460
26. Yan B, Fang NF, Zhang, PC, Shi ZH (2013) Impacts of land use change on watershed stream flow and sediment yield: An assessment using hydrologic modelling and partial least squares regression. J. Hydro 484:26-37
27. Shi Z, Ai L, Li X, Huang X, Wu G, Liao W (2013) Partial least-squares regression for linking land-cover patterns to soil erosion and sediment yield in watersheds. J. Hydrol 498:165–176
28. Li Z, Xu X, Xu C, Liu M, Wang K, Yu B (2017) Annual runoff is highly linked to precipitation extremes in karst catchments of southwest China. J. Hydrometeorol 18(10): 2745-2759
29. Wold S (1995) PLS for multivariate linear modeling. In: van der Waterbeemd H, editor. Chemometric methods in molecular design: methods and principles in medicinal chemistry. Weinheim: Chemie 195-218

Check for updates

Chapter 8
Evaluation of ArcSWAT Model for Streamflow Simulation in the Humid Tropical Netravathi Catchment

N. C. Sanjay Shekar[1] (✉) and Pathak A. Abhishek[2]

[1] Department of Civil Engineering, JSS Academy of Technical Education, Bangalore, Karnataka 570060, India
sanjayshekarnc@jssateb.ac.in

[2] Department of Civil Engineering, Dayananda Sagar College of Engineering, Bangalore, Karnataka 560078, India

Abstract. The present study was taken up to evaluate the applicability and performance of the Soil and Water Assessment Tool (SWAT) hydrological model in the humid tropical Netravathi catchment (3314 km^2) located in Karnataka State, India. The ArcSWAT model version, which is integrated into the ArcGIS platform was used. The main objective of the study was to evaluate the performance of the model in simulating streamflows when the sensitivity of the model was carried and calibrated using observed flows. Therefore, a model application involving the delineation of the catchment into sub-basins and hydrological response units (HRU) was adopted. The analysis was carried out based on the daily rainfall records from influencing rain gauges and climate records availability from nearby meteorological stations, which were used as inputs to the model for the period 2011–2016. Other inputs included a digital elevation model (DEM), land use/land cover (LU/LC) classes derived from satellite remote sensing imagery and hydrologic soil groups derived from soil map. Daily streamflow records for the main catchment outlet were used for model performance evaluation, sensitivity and calibration, respectively. Delineation of the Netravathi catchment using the information on elevation, LU/LC and soil types yielded 35 sub-basins and 136 hydrologic response units (HRU). A daily time step was adopted for model application. Sensitivity analysis of the model identified curve number (CN), soil evaporation compensation factor (ESCO) and available water holding capacity by soil (SOL_AWC) as the most sensitive model parameters. Subsequently, the model was calibrated for the period 2011–2014 by manually fine-tuning these parameters to minimize the sum of the squared differences between simulated and observed daily flows at the catchment outlet (Bantwal). The model calibrated in this manner was subject to validation for the period 2015–2016. The coefficient of determination (R^2) and Nash–Sutcliffe model efficiency (NSE) were used to evaluate the performance of the model in simulating daily streamflows during calibration and validation phases. For the catchment outlet, model validation yielded R^2 and NSE values of 0.85 and 0.82, respectively, for daily streamflow comparisons, indicating reasonably good performance as compared to the literature review. Results

R. K. Bhattacharjya et al. (eds.), *Sustainable Water Resources Management*,
Advances in Sustainability Science and Technology,
https://doi.org/10.1007/978-981-16-7535-5_8

showed that the ArcSWAT model was also able to simulate daily streamflows satisfactorily. The calibrated model was later used to simulate streamflows for the future period predictions under changed HRU conditions. Overall, it appears that the SWAT model can be adopted for hydrological analyses and water resources management.

Keywords: Streamflow · HRU · Sensitivity · Calibration · Validation

1 Introduction

The demand for water resources is increasing; hence, by better understanding the movement of water with the help of water budget equation, sustainable management of water by watershed models can be implemented. Effective methods and mechanisms should be used to keep water sustainable. In a current situation, models are used to represent the characteristics of the hydrological regime. In this study area, streamflow provides a potential resource for hydropower development, consumptive use and irrigation. Hence, it is essential to assess the amount of rainfall converted to streamflow after losses. Assessing the incoming water and outgoing water inside and outside the basin using the hydrological model was necessary to increase the sustainability of water resources [8]. The ArcSWAT is an ArcGIS extension, is a graphical user interface for the SWAT [1]. It is evolved from AVSWAT, which is an ArcView extension developed for an earlier version of SWAT [4]. ArcGIS-SWAT extracts hydrologic information from spatial data for preparing SWAT input files, runs SWAT and writes the SWAT output on the data model [9]. Water resources management and hydrologic modelling studies are intrinsically related to the spatial processes of the hydrologic cycle. Distributed parameter hydrologic Soil and Water Assessment Tool model [2] generally subdivide the watershed into smaller sub-basins. It requires more data on model inputs such as topography, soil, weather and land use that affect model parameters to capture the spatial variability of the watershed subdivided into sub-watersheds that are assumed to be homogenous. Modelling the working of the hydrological cycle at the catchment scale is an important activity being pursued by researchers and practitioners for several decades. Various types of models which simulate the major water balance components using routinely available data have been developed. Such models provide a tool for assessment of spatial and temporal variations in water availability and also provide crucial information on the hydrological impacts of various anthropogenic activities and climate change. However, no single model can be expected to work well in all hydroclimatic regimes of the world. In recent years, the SWAT model has gained universal popularity, and several cases of its application in a wide variety of hydroclimatic conditions have been presented in the literature.

However, very few studies have explored the performance of the SWAT model in humid tropical regions. The accuracy with which hydrological components can be simulated under conditions of extremely high rainfall in the presence of moist tropical evergreen forests and well-drained lateritic soils has not been investigated extensively. Also, an evaluation of the manner in which input data for the model can be obtained and pre-processed under data-scarce conditions needs to be evaluated. The main objective of the study was to explore the applicability of the ArcSWAT model to a humid tropical catchment and to evaluate the nature of input data sources/requirements necessary for routine application of the model under data-scarce conditions. The performance of the model is also assessed in terms of the accuracy with which streamflows can be simulated. In particular, the study aims to focus on a distributed implementation of the model and thereby assess its performance in simulating streamflows. The study can also explore how the SWAT model can be used to analyse the impacts of land use/land cover changes on streamflows. Therefore, the present study was taken up to evaluate the performance of the SWAT model in a catchment located in the humid tropical region extending from the West coast of India to the Western Ghats mountain ranges. The selected Netravathi catchment is located in Karnataka State, India and presents an ideal case for evaluating the performance of the SWAT model under a hydroclimatic regime unique to the West Coast region of India.

2 Study Area

The Netravathi basin (Fig. 1) extends over an area of 3314 km^2. The river originates at Bellarayadurga in the Western Ghats portion of Dakshina Kannada district at an altitude of 1000 m. It flows towards west up to its confluence with the Arabian Sea south of Mangalore city. The river basin consists of many sub-basins, namely Kumaradhara, Kallaji hole, Gowri hole, Belthangadi hole, Netravathi Hole, Neriya hole and Shisla hole. This river is the main source of drinking water to Mangalore city and surrounding areas and also provides irrigation water supplies to agricultural and plantation crops in the basin. The physiography of the basin (Fig. 1) consists of upland Western Ghats hill slopes (high elevation), the mid-land region with undulating topography (medium elevation) and the coastal plains (low elevation). Soils in the coastal plains and mid-land region are lateritic in nature, whereas red loamy soils exist on the hill slopes. Lateritic soils are well-drained and possess low water retention capacities. The geology is characterized by lateritic formations underlain by granite rocks at depths of about 50–100 m. Predominant vegetation types in the mid-land and coastal plains are deciduous trees, shrubs, mangroves, coconut and arecanut palms and patches of rain-fed paddy fields. The Western Ghats mountains shelter a wide array of non-equatorial tropical vegetation from fragments of wet evergreen to dry deciduous forest habitats in various stages of degradation to mountain forests and grasslands, alternating with zones converted into agroforests, monoculture plantations and agriculture [7]. The climate of the region according to Thornthwaite's classification [3] is humid tropical. The average annual rainfall of the region is about 4100 mm, with the bulk (93%) of the rain occurring during the months of June through September because of the southwest monsoon phenomenon. Annual rainfall varies from about 3300 mm at the coast to as high as 6000 mm in the mountains primarily due to orographic effects. The region is characterized by moderate air temperatures throughout the year (25–30 °C) and high levels of relative humidity (65–88%) (Fig. 2).

Daily records of rainfall, temperature, relative humidity, wind speed and sunshine hours are used in the application of the SWAT model which was obtained from IMD (India Meteorological Department) and Agricultural Department, Mangalore. Precipitation data from seven rain gauge stations. Temperature data is obtained from the Bajpe and

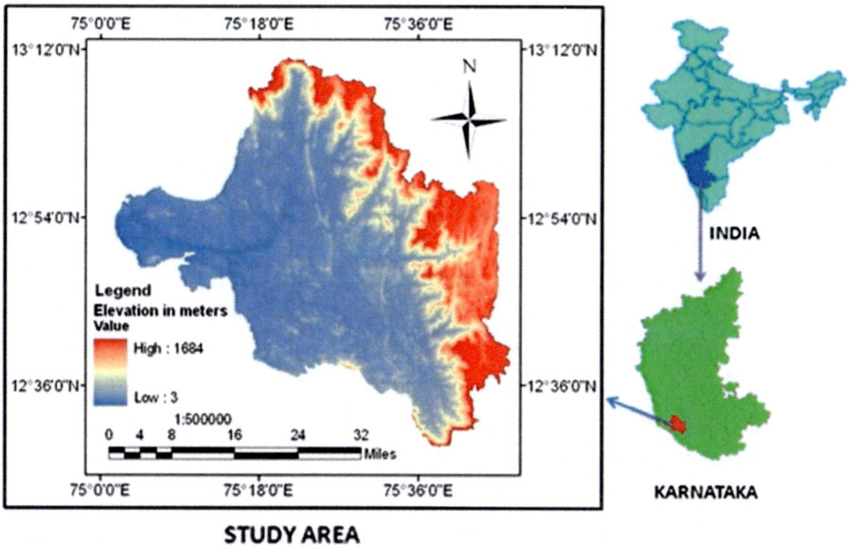

Fig. 1 Geographic location and Topography of Netravathi river basin

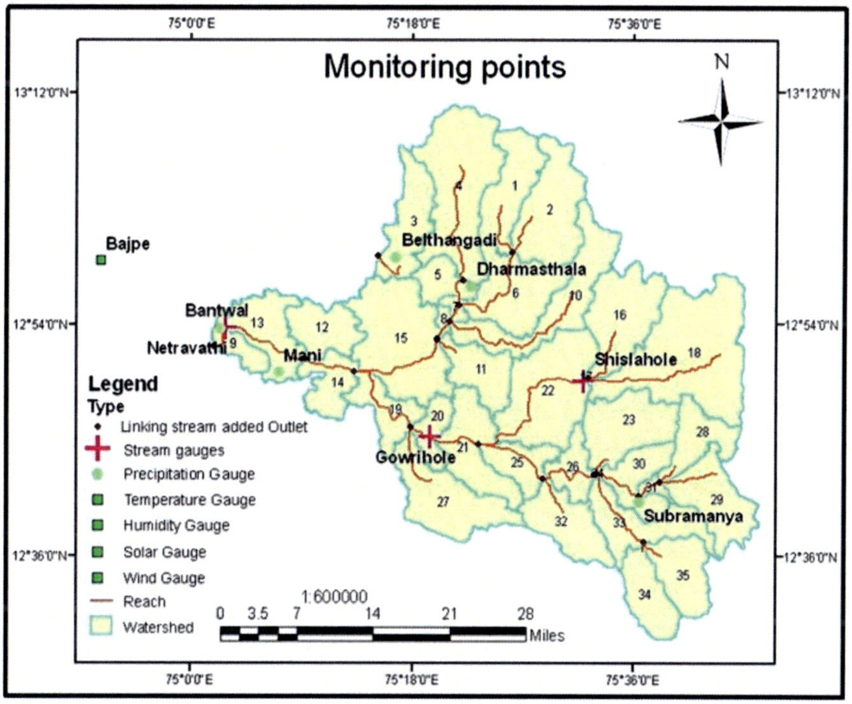

Fig. 2 Monitoring points of the Netravathi river basin

Vittal climate stations. Rain gauges/climate station details are shown in Table 1, where RF—Rainfall, AT—Average Temperature, WS—Wind Speed, RH—Relative Humidity, SSH—Sun Shine Hours. Daily streamflow data for the catchment outlet (Bantwal) is obtained from Mangalore Irrigation Department, Government of Karnataka, for a period of six years (2011–2016) is used. Soil map and database for the Netravathi basin is obtained from the NBSS and LUP (National Bureau of Soil Survey and Land Use Planning) data. Important soil parameters for 17 categories of soil types exist in the river basin are extracted from the data. LU/LC maps for the study area for the respective time period were downloaded from Bhuvan—Thematic services website. The data used in the present study is of scale 1:250,000 derived from Resourcesat-1 satellite's Linear Imaging Self-scanning Sensor LISS-III data.

Table 1 Stations in the Netravathi river basin

Sl No	Rain gauge/climate stations	Type of data used	Latitude	Longitude	Elevation (m)
1	Bantwal	RF	12° 53′ 44.81″ N	75° 02′ 25.18″ E	24
2	Bajpe	AT	12° 58′ 50.07″ N	74° 52′ 42.86″ E	26
3	Mani	RF	12° 50′ 17.19″ N	75° 07′ 12.77″ E	68
4	Vital	RF, AT	12° 57′ 15.13″ N	75° 25′ 13.22″ E	70
5	Koila	RF	12° 43′ 10.11″ N	75° 32′ 18.45″ E	90
6	Dharmasthala	RF	12° 56′ 57.12″ N	75° 22′ 56.12″ E	92
7	Puttur	RF, WS, RH, SSH	12° 45′ 34.12″ N	75° 15′ 09.12″ E	105
8	Subrahmanya	RF	12° 40′ 07.03″ N	75° 36′ 33.05″ E	110

3 Methodology

SWAT model is applied to a river basin developed by Dr. Jeff Arnold for USDA-ARS in the early 1990's. SWAT has undergone continued review and expansion of capabilities. It is a physically-based model, which uses readily available input, computationally efficient and continuous-time model. Increased accuracy in assessing the impact of land management practices in water, sediment and agricultural chemical yield in large complex watersheds. The minimum climatic inputs required by the model are maximum and minimum air temperature and precipitation. The ArcSWAT is used in the present study with ArcGIS interface as a modelling tool during streamflow validation in this research. The SWAT hydrological model requires a variety of data pertaining to climate, topography, LU/LC, soils and also streamflow records for model calibration and validation. ArcSWAT has three main components, namely watershed delineation, HRU analysis and weather data definition. For the processing of watershed delineator, DEM data and stream network is required. HRU analysis is carried out using land use, soil and slope map. Rainfall and temperature data is required for generating weather data definition for the study area. The outputs from these steps are then used as inputs for the SWAT streamflow simulation. The major components of the model are briefly discussed below.

The steps of the model setup are described below: In the first step, a new SWAT project setup is created, and then the DEM is loaded in the SWAT interface. The stream network is generated by the use of a threshold area that defines the origin of a stream. Smaller the number, more detailed the stream network generated by the interface. The locations of the different sub-basin outlets are added along with the location of the gauging site. The calibration procedure involved manual adjustment through the trial and error approach for the model parameters until the acceptable simulation is achieved. Later, sensitivity analysis is carried out to examine the relative changes in the model output with respect to change in model input variables. Sensitivity analysis also indicates the importance of the parameters in determining the streamflow in the study area. Finally, in order to utilize the calibrated model for estimating the effect of different scenarios on water balancing of the basin, the model is tested against an independent set of measured data. The model is validated with observed discharge data. The workflow diagram of the SWAT setup is shown in Fig. 3.

4 Results and Discussion

The SWAT model was applied to the Netravathi basin using the procedure described in the methodology. Based on the DEM and stream network, the catchment was delineated into 30 sub-basins. Overlay of LU/LC, soil types and DEM resulted in the identification of unique hydrological response units (HRUs). Concurrent data on required input variables and observed streamflow was available for the period 2011–2016 of which 2011–2014 was considered to be the model calibration set and data for 2015–2016 was considered for model validation. The model was implemented using a daily time step. Daily rainfall data from rain gauges and daily climate data from climatic stations were used to run the model. Daily streamflow records at the Bantwal gauging site located at the basin outlet were used for model calibration and validation. Sensitivity analysis was performed using daily observed streamflow records to identify the most critical model parameters that influenced daily runoff output. Five critical parameters identified in this manner were manually adjusted to minimize RMSE between daily simulated runoff and daily observed runoff at the gauging site. The results obtained from the study are discussed below. Following model application to the Netravathi basin with relevant inputs for the period 2011–2014, sensitivity analysis is performed, from which critical parameters: CN, ESCO and SOL_AWC are identified. These parameters are manually adjusted to achieve minimum RMSE between observed and simulated daily flows at the catchment outlet. Scatter plots of daily observed runoff at the Bantwal gauge site (outlet) with those simulated using the calibrated SWAT model are discussed below. To further evaluate the SWAT model, daily streamflow values obtained from Bantwal stream gauge station are compared with streamflow output derived from the SWAT model after calibrating using measured daily runoff available at the outlet of the catchment.

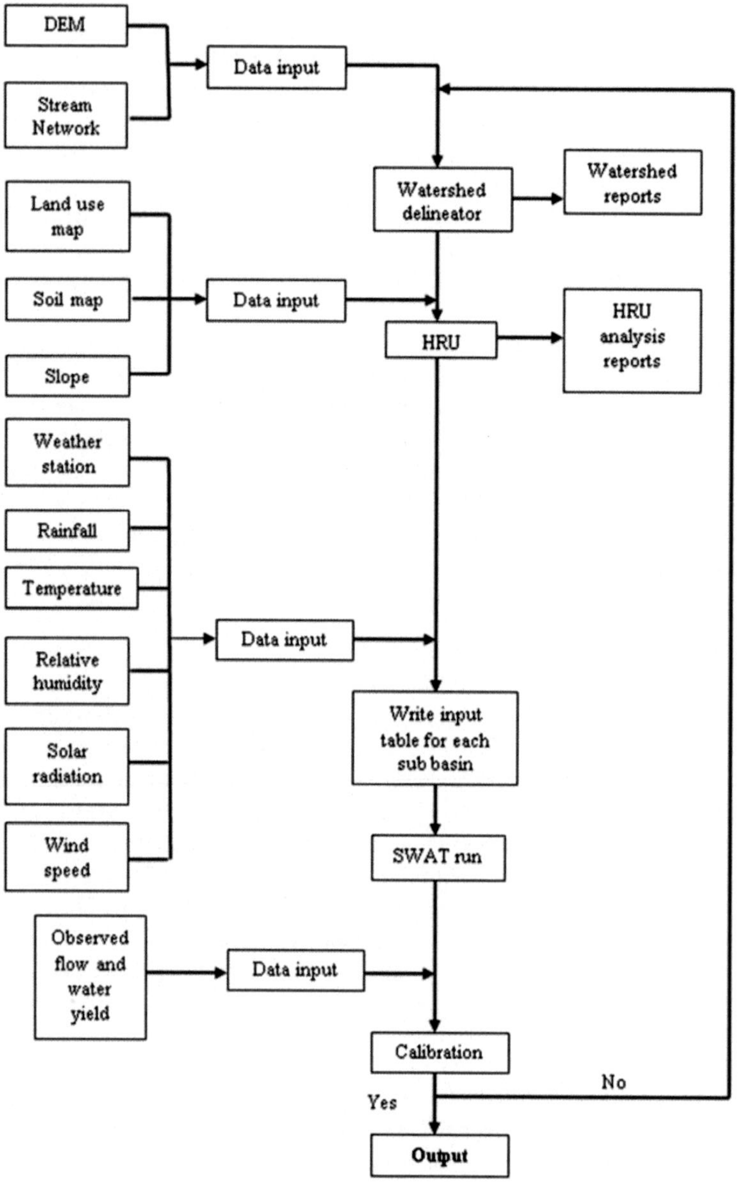

Fig. 3 Workflow diagram for setup and SWAT run

Nash–Sutcliffe efficiency (NSE) and coefficient of determination (R^2) was computed as follows

$$\text{NSE} = 1.0\sum_{t=1}^{T}\frac{(y_t - f_t)^2}{\sum_{t=1}^{T}(y_t - \overline{y})^2}$$

where y_t is the observed data values for time period t, f_t is the simulated data values for the same period, $\overline{y}$ is the mean observed data values per time period and T is the number of time periods [6]. The maximum NSE value possible is 1.0 and occurs if simulated values perfectly match observed values. The lower the NSE value, the lesser the goodness of fit between the simulated and observed time series. The larger NSE values denote better model performance.

$$R^2 = \left\{\frac{\sum_{t=1}^{T}(y_t - \overline{y})\left(f_t - \overline{f}\right)}{\left[\sum_{t=1}^{T}(y_t - \overline{y})^2\right]^{0.5}\left[\sum_{t=1}^{T}\left(f_t - \overline{f}\right)^2\right]^{0.5}}\right\}^2$$

where $\overline{y}$ is the mean of observed values for the entire evaluation time period and $\overline{f}$ is the mean of simulated values for the entire evaluation time period. The other symbols have the same meanings as defined in the preceding equation. The R^2 value is equal to the square of Pearson's product-moment correlation coefficient [5]. R^2 ranges from 0.0 to 1.0. Higher values equate to better model performance.

Model performance is evaluated by computing the coefficient of determination (R^2) and Nash–Sutcliffe model efficiency (NSE) between observed and simulated flows. From the results, it can be seen that model performance is quite good as indicated by high values of R^2 and NSE. The calibrated SWAT model is subjected to validation test using data for the period 2015–2016 and performance is evaluated using the same statistics. The calibrated model was later used to simulate streamflows under changed HRU conditions for future predictions. Results from Figs. 4 and 5 show that R^2 and NSE values are 0.81 and 0.76 before calibration (2011–2014) and 0.85 and 0.82 during the validation period (2015–2016).

Figures 6 and 7 depict that the time series hydrograph of the daily time step discharge variation between the measured and the simulated flow. The graph indicates that the maximum flow had occurred in the monsoon season. During this season, the variation of the discharge was found to range between 2000 and 4000 cumecs. The overall performance of the model was found to be satisfactory, as most of the simulated values match the measured values. Graphs clearly indicate that the sharp peak occurred every year during the monsoon period. It can be observed that the simulated values were overestimated than the measured values in this station.

5 Conclusion

Based on the modelling exercise carried out, ArcSWAT version of the model provided a suitable environment in ArcGIS for implementation. Input data compilation, preliminary

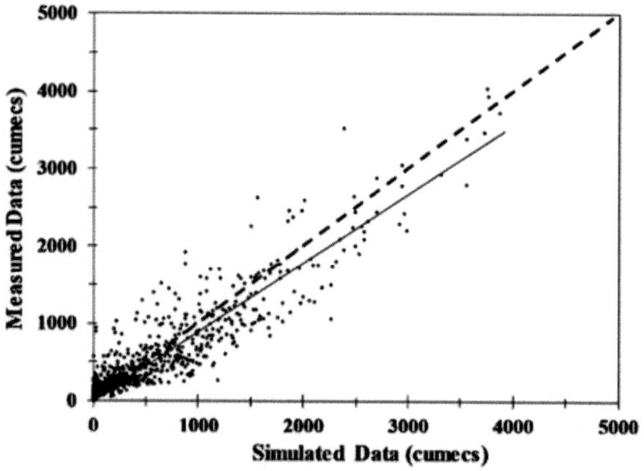

Fig. 4 Daily measured versus observed streamflow before calibration

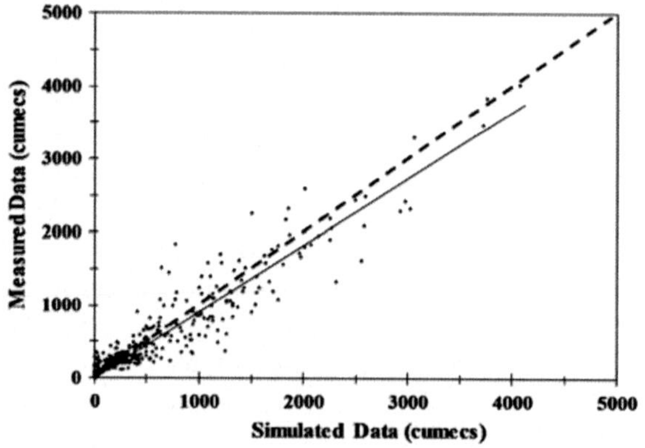

Fig. 5 Daily measured versus observed streamflow after calibration

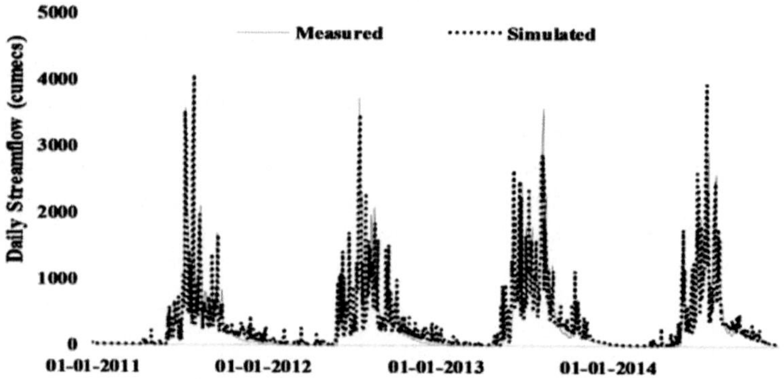

Fig. 6 Comparison of daily measured versus observed streamflow before calibration

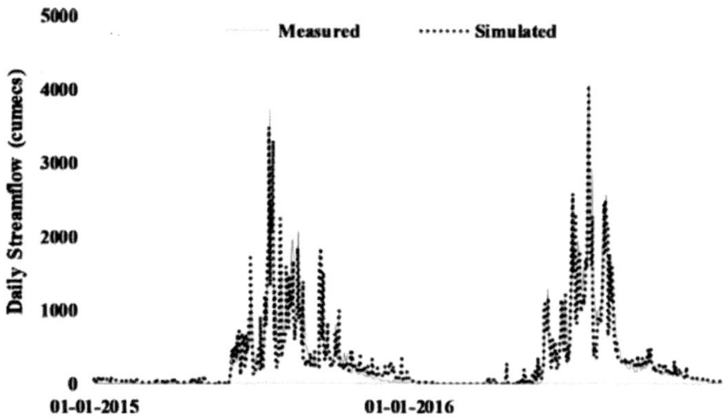

Fig. 7 Comparison of daily measured versus observed streamflow after calibration

analysis and pre-processing could be accomplished effectively and accurately. Similarly, model application and extraction of output at different time steps and spatial scales and its post-processing was equally convenient. Based on the reasonably good performance statistics of the model, it can be said that input data obtained LU/LC, DEM and soil map were accurate enough. The use of these data sources is therefore recommended for future hydrological analyses in the region. Since a distributed modelling approach was adopted in this study, ArcSWAT was very useful in automatically delineating the catchment into 35 sub-basins and 136 HRUs. Sensitivity analysis of the model showed that CN, ESCO and SOL_AWC were the most sensitive parameters with regard to streamflow estimation in the Netravathi catchment. Model performance in simulating temporal variations in streamflow of the outlet of the Netravathi river basin (Bantwal gauging station) was quite satisfactory. During the calibration period (2011–2014), R^2 value is 0.76 for daily comparisons with observed flows. The values of NSE is 0.81 for daily comparison during the calibration period. The model performance was quite good even in the validation period (2015–2016) for the catchment outlet. During this phase, the value of R^2 was 0.85 for daily comparisons with observed flows comparisons. The value of NSE is 0.82 for daily comparison, respectively. The ability of SWAT to simulate the impacts of LU/LC changes on runoff variability was studied by comparing flows for different LU/LC categories for the respective year. As a result of the decrease in forest cover and the expansion of agricultural land and urban areas, the model realistically predicted reduced streamflows. Because of the results obtained in this study, it may be concluded that ArcSWAT is a useful modelling tool for hydrologic analyses and water resources planning and management in humid tropical basins.

Conflict of Interest. None.

References

1. Arnold JG, Srinivasan R, Muttiah RS, Williams JR (1998) Large-area hydrologic modeling and assessment: part I. Model development. J Am Water Resour Assoc 34(1):73–89
2. Arnold JG, Moriasi DN, Gassman PW, Abbaspour KC, White MJ, Srinivasan R et al (2012) SWAT: model use, calibration, and validation. Trans ASABE 55(4):1491–1508
3. Cunha AR, Schoffel ER (2011) Evapotranspiration—from measurements to agricultural and environmental applications, Chapter 20: The evapotranspiration in climate classification. InTech, Chapters published, São Paulo State University—Superior School Agronomic Sciences Federal University of Pelotas—Superior School Eliseu Maciel Agronomy, Brazil, pp 391–410
4. Di Luzio M, Srinivasan MR, Arnold J (2001) ArcView interface for SWAT 2000-User's guide. USDA agricultural research service, Temple, Texas
5. Legates DR, McCabe GJ (1999) Evaluating the use of "goodness-of-fit" measures in hydrologic and hydroclimatic model validation. Water Resour Res 35(1):233–241
6. Nash JE, Sutcliffe JV (1970) River flow forecasting through conceptual models: part 1. A discussion of principles. J Hydrol 10(3):282–290
7. Renard Q, Pelissier R, Ramesh BR (2010) Environmental susceptibility model for predicting forest fire occurrence in the Western Ghats of India. Int J Wildland Fire 21(4):368–379
8. Vijverberg J, Dejen E, Sibbing FA (2009) Lake Tana: source of the Blue Nile series: Monographiae Biologicae 89:164–192. ISBN: 978-1-4020-9725-6
9. Olivera F, Valenzuela M, Srinivasan R, Choi J, Cho H, Koka S, Agrawal A (2006) ArcGIS-SWAT: a geodata model and GIS interface for SWAT. J Am Water Resour Assoc 42(2):295–309

Chapter 9
Parameter Sensitivity by Watershed Model of Bhogdoi, a Tributary of the Brahmaputra River

Monisha Dutta(✉), Pulendra Dutta, and Mrinal Kumar Dutta

Civil Engineering Department, Jorhat Engineering College, Jorhat 785007, India

Abstract. Water being one of the most critical natural resources needs efficient and sustainable management in the scenario of rapid urbanization and development. Hydrologic models prove to be an effective technique in this regard. An understanding of the watershed scale hydrologic processes at the spatial level stands in need of geospatial and hydro-meteorological data, represented by mathematical equations. A hydrologic model established using hydro-meteorological data at a large spatial scale for a river basin is not expected to provide better results. Therefore, several sets of these data are utilized while hydrological modeling of the Bhogdoi River Basin in the Soil and Water Assessment Tool (SWAT) platform. The main objective of this study is to identify various parameters sensitive to the river basin for providing better simulation results. Here, a global and local sensitivity analysis using the SUFI-2 (sequential uncertainty fitting ver.2) algorithm was carried out to review the linkage between the input and output variables in hydrological simulation.

Keywords: Hydrologic processes · Sensitivity · SWAT · SUFI-2

1 Introduction

Hydrological parameters such as precipitation, surface runoff, infiltration, evaporation, river discharge, soil moisture, etc., play a vital role in the representation of the hydrological cycle, the knowledge of which is widely beneficial for water resource management. Precipitation and discharge are the two key parameters that are incorporated in hydrological models to study the occurrence of hydrological hazards such as floods, droughts, etc., and the performance of hydraulic structures. SWAT (Soil and Water Assessment Tool), developed by the U.S. Department of Agriculture is a continuous-time model used to simulate at the basin scale hydrological processes and is widely used by researchers all around the world to perform scenario analysis in terms of flood risk assessment [7], land use and climate change impact [2], management practices [8].

SWAT-CUP, a Calibration Uncertainty Program is used to analyze the uncertainty of the calibration and validation results of the SWAT model by linking SUFI-2, particle

R. K. Bhattacharjya et al. (eds.), *Sustainable Water Resources Management*, Advances in Sustainability Science and Technology,
https://doi.org/10.1007/978-981-16-7535-5_9

swarm optimization, generalized likelihood uncertainty estimation (GLUE), parameter solution (Parasol) and Markov Chain Monte Carlo (MCMC) algorithms to SWAT model. SUFI-2 algorithm predicts all uncertainties (input data, parameters, model structure, output data) by finding the best amount of parameter uncertainty, the uncertainty being determined by the 95% prediction uncertainty band calculated at 2.5% and 97.5% levels of the output variables. The p-factor is the percentage of measured data bracketed by the 95% prediction boundary (95 PPU) and the r-factor quantifies the strength of a calibration/ uncertainty analysis. A p-factor and r-factor of 1 and 0, respectively, correspond exactly to the measured data as suggested [1].

The study focuses on the sensitivity analysis of the selected parameters responding to streamflow in the Bhogdoi Basin based on global and local sensitivity analysis. Special highlights were given on the parameters that contribute to the better results of the selected objective function.

2 Materials and Methods

Study Area

The Bhogdoi River Basin lying between 26° 16′ 40″ and 26° 49′ 10″ north lines of latitudes and 94° 2′ 50″ and 94° 28′ 50″ east lines of longitudes with the river originating at Long Samtang of Mokukchang, Nagaland and discharging at Kakodonga River in northwest of Jorhat was selected for the present study (Fig. 1). The study area covers an area of 833.6 km^2 and demonstrates a peculiar V-shape, thus having a squeezed central portion with the width gradually increasing towards the north and the south.

Input Data

The SRTM (shuttle radar topographic mission) digital elevation model (DEM) selected for the study having a spatial resolution of 30 m was obtained from opentopography.org. A 100 m resolution LULC (Land use/Land cover) map of India was obtained from daac.ornl.gov and a 1:5,000,000 scale world map from FAO/UNESCO Soil Map of the world. Meteorological datasets from 1985 to 2019 for Jorhat was obtained from the Assam Agriculture University, Jorhat, and the discharge data at the Bhogdoi A.T. Road Crossing available for the years 1985–2000 were obtained from the Upper Investigation, Water Resource Department, Jorhat.

Model Setup and Simulation

The DEM was made free from sinks to preserve the characteristics of the DEM [6] and used for watershed delineation and stream network generation using the 'Watershed Delineator Tool' of ArcSWAT which is based on the eight pour point algorithm or D8 algorithm [5]. A threshold value of 3000 hectares was used as a threshold area based on comparison with the real stream network to create stream network and subwatershed outlets. The downloaded LULC and soil maps were clipped to the watershed boundary and multiple slope classes approach was considered for the HRU (hydrologic response

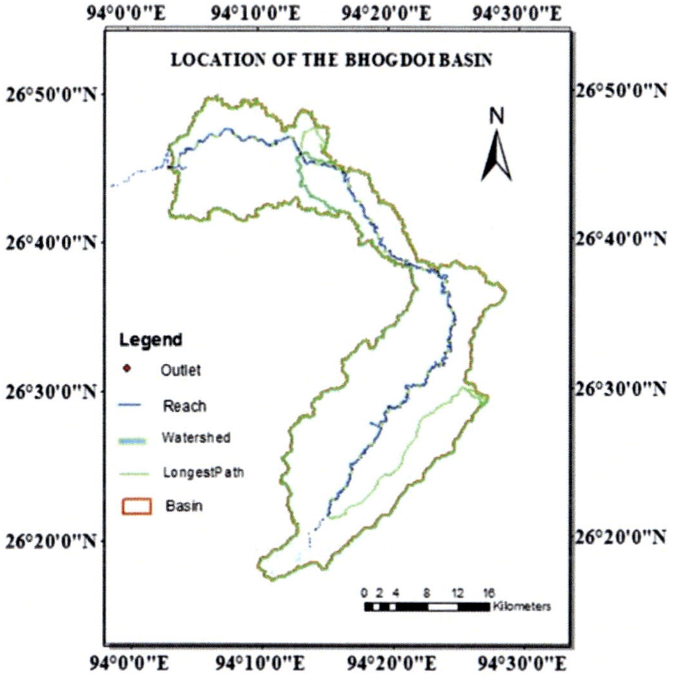

Fig. 1 Location of the study area

unit) analysis. HRUs lump similar land uses soils and slopes within a particular sub-basin depending upon the threshold provided by the user. The slope classes and their upper limits were fixed depending upon the classification (natural breaks) in the slope map of the DEM created in GIS (geographic information system). The delineated basin was subdivided into two sub-basins and 28 HRUs. Daily precipitation and maximum and minimum temperature were used in the simulation of discharge and the model was run on a monthly time step with a warm-up period of two years. SWAT model simulates the hydrological cycle based on the water balance equation given by Eq. (1) [4] which acts as the main driving principle behind every process in the SWAT model [3].

$$\mathrm{SW_t} = \mathrm{SW_o} + \sum_{i=1}^{t} \left(R_{\mathrm{day}} - Q_{\mathrm{surf}} - E_{\mathrm{a}} - W_{\mathrm{seep}} - Q_{\mathrm{gw}} \right) \tag{1}$$

where $\mathrm{SW_t}$ is the final soil water content (mm), $\mathrm{SW_o}$ is the initial soil water content in the day i (mm), t is the time (days), R_{day} is the measure of precipitation in the day i (mm), Q_{surf} is the measure of surface runoff in the day i (mm), E_{a} is the amount of evapotranspiration in the day i (mm), W_{seep} is the measure of water entering the vadose zone from the soil profile on the day i (mm) and Q_{gw} is the measure of groundwater discharge in the day i (mm). The observed discharge was compared with the simulated discharge obtained as a result of default model simulation which presented the need for parameter adjustment for better accuracy. This led to the optimization of parameters using sensitivity analysis.

Global and Local Sensitivity Analysis

After the simulation results were successfully linked to the SWAT CUP file, the objective function was selected to be the coefficient of determination R^2. Later, the sensitivity analysis was conducted during the calibration period (1985–1997) with two years as the warm-up period.

3 Results and Discussion

Results on Sensitivity Analysis

Twenty parameters were selected for sensitivity analysis by watershed model of Bhogdoi. Then, the local one-at-a-time sensitivity analysis was done using the parameter ranges as shown in Table 1 which was based on the ranges within which the parameter values in the SWAT project lies. Local sensitivity analysis examines the change in model output due to one parameter when the other parameters are constant. In the next step, a global sensitivity analysis was conducted for the same parameters within the same ranges. The global sensitivity considers the contribution of different model parameters to the model output uncertainty.

From Table 2, it can be clearly seen that the most sensitive parameter CN2 in global sensitivity analysis gets a rank of 4 in local sensitivity which indicates the fact that global sensitivity shows better results due to interaction between the other parameters. Also, parameters like SOL_ALB, SHALLST, GWQMN, DEEPST, and SOL_BD can be neglected due to their low sensitivity in both the analysis.

Calibration Results

The local sensitivity analysis carried out for the selected 20 parameters one-at-a-time produced an R^2 value of around 0.42 for all the parameters, with 5 iterations for each parameter. Therefore, the most sensitive parameter of local sensitivity analysis (RCHRG_DP.gw) was considered and the scatter plot was drawn as shown in Fig. 2a. The global sensitivity analysis carried out for the same parameters as selected in local sensitivity analysis produced R^2 value as 0.45 for 800 iterations (Fig. 2b). Owing to the better performance of global sensitivity analysis, it was considered to check the remarkable difference between the observed and simulated discharge. It is visible from Fig. 3 that as we have adjusted the most sensitive parameter CN2 for each HRU with other parameter ranges kept the same as before; it significantly reduced the surface runoff, thereby increasing the R^2 value to 0.48. In this process, the parameters were provided to SWAT-CUP in decreasing order of sensitivity and 500 iterations were given. Thus, the sensitive parameters need to be adjusted to reduce this difference and improve R^2 above the satisfactory level of 0.5. The validation can be then carried out for the best-fitted parameter ranges for the remaining years with available discharge data and discharge can be simulated for the most recent years.

Table 1 Maximum and minimum parameter ranges in the sensitivity analysis

No	Parameter	Parameter description	Ranges	
			Min	Max
1	r_CN2.mgt	SCS runoff curve number *f*	-0.1	0.05
2	v_ALPHA_BF.gw	Base flow alpha factor (days)	0	0.5
3	v_SURLAG.hru	Surface runoff lag time (days)	1	5
4	v_SLSUBBSN.hru	Average slope length	80	120
5	r_USLE_K.sol	USLE equation soil erodibility (k) factor	-0.2	0.2
6	v_REVAPMN.gw	Threshold depth of water in the shallow aquifer for revap to occur (mm)	500	1000
7	r_SOL_K.sol	Saturated hydraulic conductivity	-0.2	0.2
8	v_GW_DELAY.gw	Groundwater delay (days)	10	50
9	v_HRU_SLP.hru	Average slope steepness	0	0.6
10	v_RCHRG_DP.gw	Deep aquifer percolation factor	0	0.1
11	v_EPCO.hru	Plant uptake compensation factor	0.5	1
12	r_SOL_ALB.sol	Moist soil albedo	-0.2	0.2
13	v_ESCO.hru	Soil evaporation compensation factor	0.5	1
14	v_OV_N.hru	Manning's '*n*' value for overland flow	0.05	0.5
15	a_SHALLST.gw	Initial depth of water in the shallow aquifer (mm)	-500	500
16	v_GW_REVAP.gw	Groundwater revap coefficient	0.02	0.1
17	a_GWQMN.gw	Threshold depth of water in the shallow aquifer required for return flow to occur (mm)	-500	500
18	a_DEEPST.gw	Initial depth of water in the deep aquifer (mm)	-1000	1000
19	r_SOL_BD.sol	Moist bulk density	-0.2	0.2
20	r_SOL_AWC.sol	Available water capacity of the soil layer	-0.1	0.1

Where v_ means replace an existing value with a given value, r_ means a relative change to a given value and a_ means a particular value is added to the given value

Table 2 Ranking of the parameters in the global and local sensitivity analysis

No	Input parameter	Global sensitivity			Local sensitivity		
		p-stat	t-stat	Ranking	p-stat	t-stat	Ranking
1	r_CN2.mgt	0	-11.04	1	0	-24.50	4
2	v_ALPHA_BF.gw	0	5.11	2	0	-6.10	8
3	v_SURLAG.hru	0	-2.59	3	0.01	-5.19	10
4	v_SLSUBBSN.hru	0.06	-1.82	4	0.01	-4.89	11
5	r_USLE_K.sol	0.08	-1.73	5	0.06	2.93	14
6	v_REVAPMN.gw	0.08	1.69	6	0	68.70	2
7	r_SOL_K.sol	0.14	1.46	7	0.68	-0.45	18
8	v_GW_DELAY.gw	0.14	-1.45	8	0.04	-5.52	9
9	v_HRU_SLP.hru	0.15	-1.43	9	0	-62.41	3
10	v_RCHRG_DP.gw	0.15	-1.41	10	0	-226.52	1
11	v_EPCO.hru	0.20	1.27	11	0.54	-0.67	17
12	r_SOL_ALB.sol	0.24	1.16	12	1	0	20
13	v_ESCO.hru	0.29	1.05	13	0.09	2.45	15
14	v_OV_N.hru	0.46	-0.72	14	0	17.70	6
15	a_SHALLST.gw	0.49	0.67	15	0.05	-2.98	13
16	v_GW_REVAP.gw	0.72	-0.35	16	0	68.70	2
17	a_GWQMN.gw	0.83	-0.20	17	0.06	2.93	14
18	a_DEEPST.gw	0.84	0.20	18	0	0	19
19	r_SOL_BD.sol	0.84	0.20	19	0.03	3.53	12
20	r_SOL_AWC.sol	0.94	-0.07	20	0	7.99	7

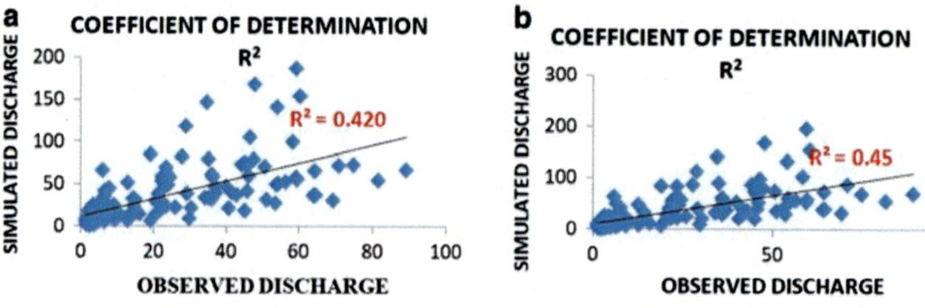

Fig. 2 Scatter Plot for **a** Local sensitivity analysis (RCHRG_DP.gw). **b** Global

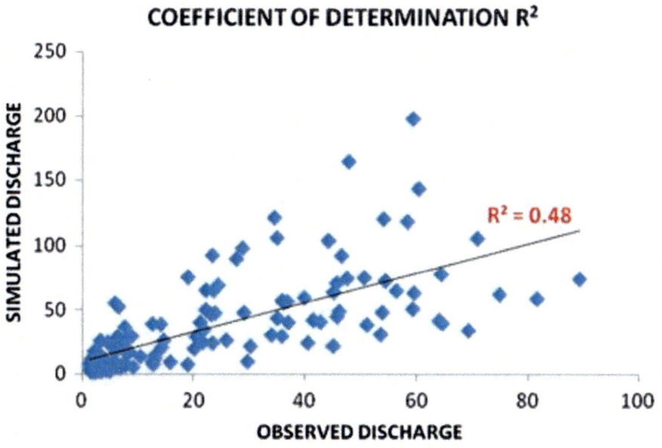

Fig. 3 Scatter Plot after parameter adjustment (CN2) in the global sensitivity analysis

4 Conclusion

The SUFI-2 algorithm in SWAT CUP proved beneficial for the selection of the model parameters most sensitive to the model output thus reducing the model uncertainty. Here, it was observed that the global sensitivity analysis seemed to improve the simulation results better. Thus, global sensitivity can be used along with the adjustment of the sensitive parameters for the calibration and validation required for further scenario analysis.

Acknowledgements. The Assam Agriculture University, Jorhat and The Upper Investigation Department, Water Resource Department, Jorhat are acknowledged for providing the hydro-meteorological datasets useful for the project.

References

1. Abbaspour KC, Johnson CA, Genuchten M (2004) Estimating uncertain flow and Transport parameters using a sequential uncertainty fitting procedure. Vadose Zone J 1340–1352
2. Anand V, Oinam B (2019) Future climate change impact on hydrological regime of river basin using SWAT model. Glob J Environ Sci Manage 471–484
3. Arnold JG, Kiniry JR, Srinivasan R, Williams JR, Haney EB, Neitsch SL (2011) Soil and water assessment tool input-output file documentation
4. Neitsch SL, Arnold JG, Kiniry JR, Williams JR (2005) Soil and Water Assessment Tool—theoretical documentation, version 2005. Texas, USA
5. O' Callaghan JF (1984) The extraction of drainage networks from digital elevation data. Comput Vis Graph Image Process 28:323–344
6. Scarnama G (2014) Lossless data compression of grid-based digital elevation models: a PNG image format evaluation. ISPRS Ann Photogrammetry Remote Sens Spat Inf Sci II-5:23–25
7. Sufiyan I, Magaji JI (2018) Modeling flood hazard using SWAT and 3D analysis in Terengannu Watershed. J Clean WAS 2:19–24
8. Vidula S, Sushama K (2014) Evaluation of the best management practices using SWAT model for the Kaneri micro-watershed. Southern Maharashtra, India

Chapter 10
A 2D Hydrodynamic Model Study in Brahmaputra River for Implementation of Bank Protection Work at Nimatighat

Anupal Baruah[1](✉), Priyam Deka[1], Ranjit Deka[2,3], and Arup Kumar Sarma[1]

[1] Department of Civil Engineering, Indian Institute of Technology, Guwahati, India
aks@iitg.ac.in
[2] NEHARI, Brahmaputra Board, Guwahati, India
[3] IIT, Guwahati, India

Abstract. The morphological characteristics in a natural channel are widely affected by its sediment transport capacity. There are various possible reasons which may imbalance the sediment supply and transport rates and alters the flow dynamics in the channel. River Brahmaputra is associated with the problems of bank erosion, deposition of sediments, rapid bed aggradations, and drainage congestions and flooding from decades. Channel dredging is a possible method that can be implemented in the vulnerable reaches to remove the accumulated sediments and provide sufficient flow depth in the channel for navigation as well as for the safe passage of flood waves. Dredging in the channel improves the bed slope, increases the flow area and results in quick disposal of water during the high flow period. In this work, river hydraulics near the eroded bank is evaluated using a two-dimensional hydrodynamic model by considering three possible dredging widths in the channel along with a series of permeable spurs. The suitable alternative out of the proposed methods is selected based upon the computed parameters.

Keywords: Dredging · Permeable spurs · Two dimensional hydrodynamic model

1 Introduction

A river is an integrated system of water and sediments carried along with it. Any alterations in the atmospheric and terrestrial systems of a watershed are integrated and manifested in the river system. As the river flows, the dynamics of the exchange between its water and sediment load along with the geology and the hydrology of the landscape creates a complex network of processes, resulting in a given physical form of the river system. River morphodynamics is a consequence of channel dimensions, gradients and channel adjustment by an erosion deposition process [1]. The alluvial rivers are characterized by the sediment transported from its bed and bank. Most of the alluvial rivers in

R. K. Bhattacharjya et al. (eds.), *Sustainable Water Resources Management*, Advances in Sustainability Science and Technology,
https://doi.org/10.1007/978-981-16-7535-5_10

India like Ganga and Brahmaputra show erratic behaviour both during the lean period and flow period. The planform of these alluvial rivers are constantly changing and are extremely dynamic character. Equilibrium can be at best a statistical phenomenon since there must be local erosion and deposition as flow changes [2]. River channel behaviour often needs to be studied for its natural state and responses to human activities.

High silt carrying capacity and gradient variations of the Brahmaputra River causes aggradations and degradations at different locations. To prevent bank erosion in Majuli, four numbers of spurs are constructed to deflect the flow from the bank promoting the silting in the neighbouring areas. The main channel is getting contracted near the Nimatighat in the presence of a large sandbar. During monsoon, 30,000–50,000 cumec of water passes through the contracted portion causing heavy erosion near the bank. It is the interest of the stake holders to carry out a river training work in that stretch to reduce the soil and property loss. Before implementing any possible river training measures in field, a numerical or field model study is essential for the decision-making process. Proper representation of the braided stretch and bed-level variations is difficult to model in field scale, and therefore, a 2D hydrodynamic model is a good alternative for flow prediction. Hydrodynamic models are helpful in quick simulation with different flow scenarios. In this study, BRAHMA-2D model is used for simulation. This model is developed by IIT-Guwahati in collaboration with Brahmaputra Board. The basic objective of this simulation is to apply the 2D hydrodynamic model in a highly braided stream to quantify the velocity distributions near the eroded bank and to adopt the best alternative out of a single measure or from a couple of measures. Three possible dredging widths and a series of permeable spur are considered in the simulation. Morphological changes are not taken into account because the current version of BRAHMA-2D model solves the two-dimensional shallow water equation, and the sediment transport simulation is not yet incorporated.

2 Governing Equations and Numerical Scheme

Two-dimensional depth averaged shallow water equation by neglecting the viscous and Coriolis effect can be expressed in matrix form as follows [3]

$$\frac{\mathrm{d}U}{\mathrm{d}t} + \frac{\mathrm{d}F}{\mathrm{d}x} + \frac{\mathrm{d}G}{\mathrm{d}y} = S \tag{1}$$

where,

$$U = \begin{Bmatrix} \eta \\ q_x \\ q_y \end{Bmatrix} F = \begin{Bmatrix} q_x \\ {q_x}^2/h \\ q_x q_y/h \end{Bmatrix} G = \begin{Bmatrix} q_y \\ q_x q_y/h \\ {q_y}^2/h \end{Bmatrix} S = \begin{Bmatrix} 0 \\ -gh\frac{\mathrm{d}\eta}{\partial x} - s_{fx} \\ -gh\frac{\mathrm{d}\eta}{\partial y} - s_{fy} \end{Bmatrix}$$

where η is the water surface elevation (m), q_x and q_y discharge/width in x and y direction, $\frac{\mathrm{d}\eta}{\partial x}$ and $\frac{\mathrm{d}\eta}{\partial y}$ are the water surface slopes in x and y direction, s_{fx} and s_{fy} are the friction slopes in x and y direction, respectively. After transforming the equations in boundary-fitted system [4]

$$\frac{\partial}{\partial t}[J\eta] + \frac{\partial}{\partial \xi}[JhU] + \frac{\partial}{\partial \eta}[JhV] = 0 \tag{2}$$

$$\frac{\partial}{\partial t}[Jhu] + \frac{\partial}{\partial \xi}[J\{huU\}] + \frac{\partial}{\partial \eta}[J\{huV\}] = -Jgh\left(\frac{\mathrm{d}\eta}{\mathrm{d}x} - s_{fx}\right) \tag{3}$$

$$\frac{\partial}{\partial t}[Jhv] + \frac{\partial}{\partial \xi}[J\{hUv\}] + \frac{\partial}{\partial \eta}[J\{hVv\}] == -Jgh\left(\frac{\mathrm{d}\eta}{\mathrm{d}y} - s_{fy}\right) \tag{4}$$

where,

$$U = \left(\frac{\partial \xi}{\partial x}u + \frac{\partial \xi}{\partial y}v\right) \text{and} V = \left(\frac{\partial \eta}{\partial x}u + \frac{\partial \eta}{\partial y}v\right), J = \begin{vmatrix} \frac{\partial x}{\partial \xi} & \frac{\partial x}{\partial \eta} \\ \frac{\partial y}{\partial \xi} & \frac{\partial y}{\partial \eta} \end{vmatrix} = \begin{vmatrix} \frac{\partial x}{\partial \xi} & \frac{\partial y}{\partial \xi} \\ \frac{\partial x}{\partial \eta} & \frac{\partial y}{\partial \eta} \end{vmatrix}$$

A second-order accurate explicit Mc-Cormack predictor–corrector scheme is used for the solution of the unsteady flow equations. A splitting algorithm is employed for the solution in which the two-dimensional equations are divided into four one-dimensional equations, and each 1D equations are solved successively. Variables determined in the predictor part are used during the corrector part. The non-physical oscillations near the steep regions need to be smoothened by some specific numerical treatment, otherwise this oscillation propagates in the solution leading to the failure of the model. Addition of coefficient of artificial viscosity in the solution is one technique to suppress these dispersions, but it requires a trial and error process to select the proper amount of diffusion [5]. TVD scheme is a method that can be used to add the right amount of dissipation subsequently at the sharp region. The benefit of the TVD model is that in this scheme, the proper amount of weightage to be added at the steep regions is decided algebraically, and it is also free of adjustable parameters. Five points total variation diminishing scheme is used in this model [6]. Stability of the model is ensured by CFL condition.

3 Study Area of the Proposed Dredging for Modelling Exercise

Majuli, the nerve-centre of the Neo-Vaishnavite culture, is situated between 26° 45′ N–27° 12′ N latitude and 93° 39′ E–94° 35′ E longitude. Map of the study area is shown in Fig. 1a. It is seen from Fig. 1b that due to the presence of series of spurs at the upstream of the reach deflect the flow towards the south bank. Since the river channel is contracting from 1.30 km upstream to 730 m downstream, and when the flow takes place through this contracted curvature part with high velocity, it causes erosion near the bank. Flow area near the affected bank can be increased by dredging some portion of the sandbar and diverting some amount of water through the dredged channel. A combination of permeable spurs at the upstream of the dredged channel is another alternative that can be implemented. Prior to the implementation, feasibility of the proposed suggestions for channel improvement is studied by employing the mathematical model.

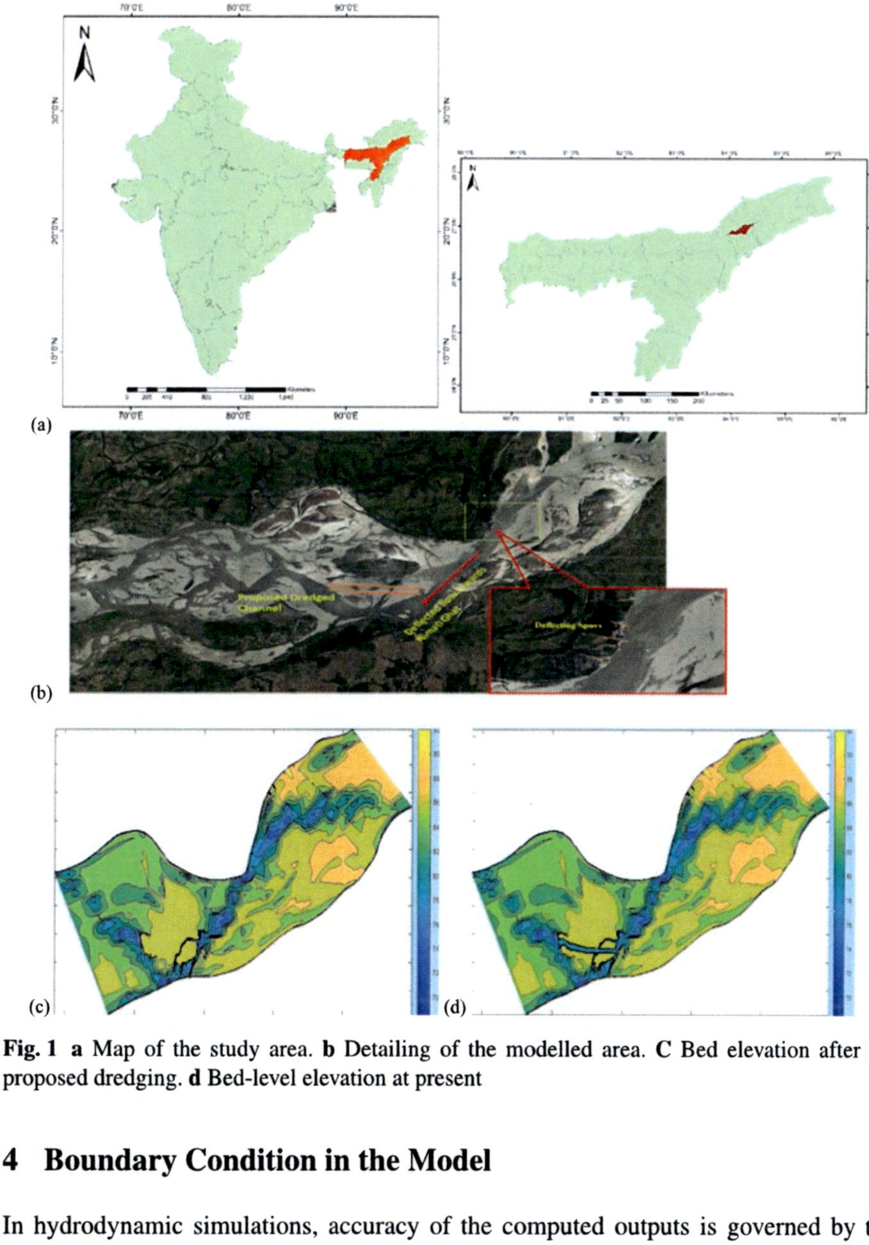

Fig. 1 **a** Map of the study area. **b** Detailing of the modelled area. **C** Bed elevation after the proposed dredging. **d** Bed-level elevation at present

4 Boundary Condition in the Model

In hydrodynamic simulations, accuracy of the computed outputs is governed by the boundary conditions, especially at the upstream and downstream location. The model simulations with incorrect datasets increase the uncertainty in the model and thus the decision-making processes. Data scarcity is a big issue in model simulation. Hydrodynamic models with wrong boundary conditions leads to an inferior solution. In this model, a stage-discharge curve for different return period is prepared from the frequency analysis of the hydrological datasets of 20 years collected from different sources (Fig. 2).

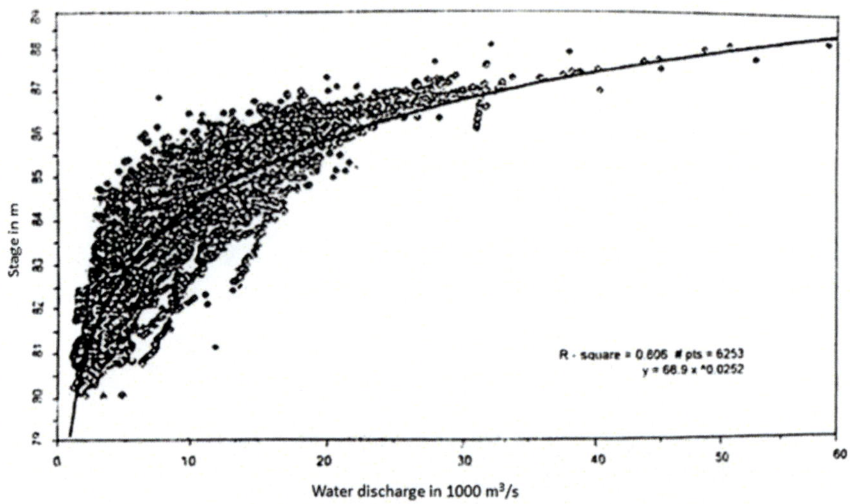

Fig. 2 Rating curve near Majuli. Courtesy: Brahmaputra Board

5 Result and Discussion

First Case: Dredging of the Sandbar with Three Different Width

In the first case, model is simulated by considering the three different dredging widths in the channel. Width in probable dredged locations is considered as 150, 300, and 450 m.

Current speed near the banks and along the centreline the dredged channel is plotted in Fig. 3b, c. From this simulation, it is observed that out of the three possible dredging conditions considered here, channel having dredged width 450 m considerably reduces the near bank velocity and increases the current speed within the channel as compared to the others.

Second Case: Dredging of the Sandbar with Porcupine Screen

From the above simulation, it is found that with 450 m dredging width, there is a considerable reduction in velocity profile. In the second simulation, a series of porcupine screen is installed from the upstream of the dredged channel and extended up to the affected portion of the bank. Figure 4a shows the locations for installation of the porcupine screens.

In the hydrodynamic model, porcupine screens are simulated as permeable spurs. Permeable spurs are used to reduce the velocity and promote siltation but not to deflect the flow. In simulation, roughness values at those points are increased to modify the momentum fluxes. Velocity profiles are plotted near the bank.

From Fig. 4b, by comparing the computed outputs under two scenarios, it is seen that there is not any significant reduction in current speed near the bank after installing the porcupine screens at the upstream along with the 450 m dredging.

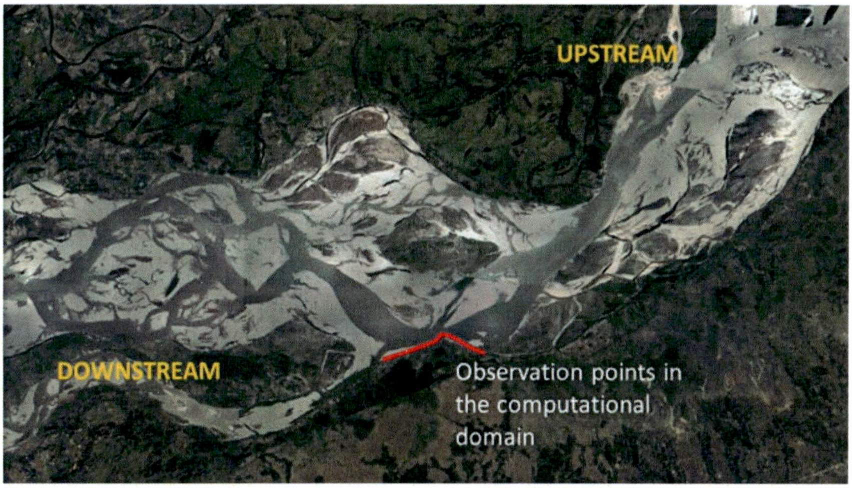

(a) Flow parameter observation points near the south bank

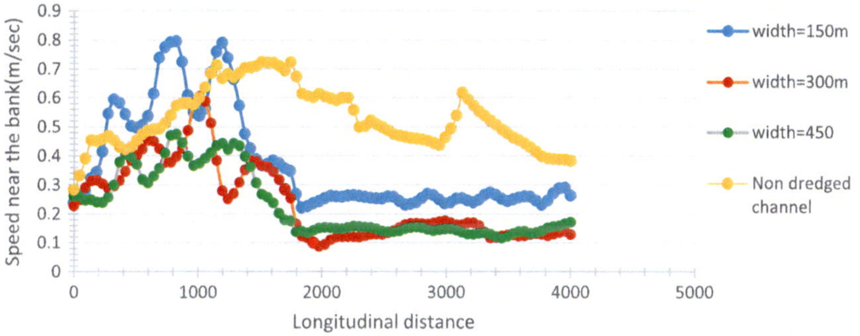

(b) Water speed near the south bank at different dredging width

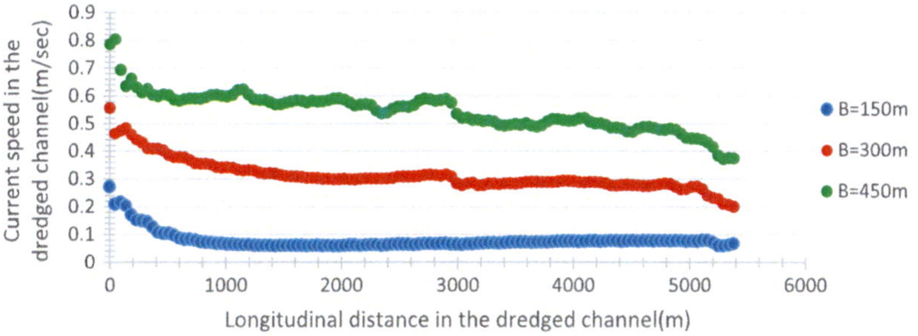

(c) Water speed in the dredged channel at different dredging width

Fig. 3 **a** Flow parameter observation points near the south bank. **b** Water speed near the south bank at different dredging width. **c** Water speed in the dredged channel at different dredging width

(a) Proposed location for RCC porcupine screen installation

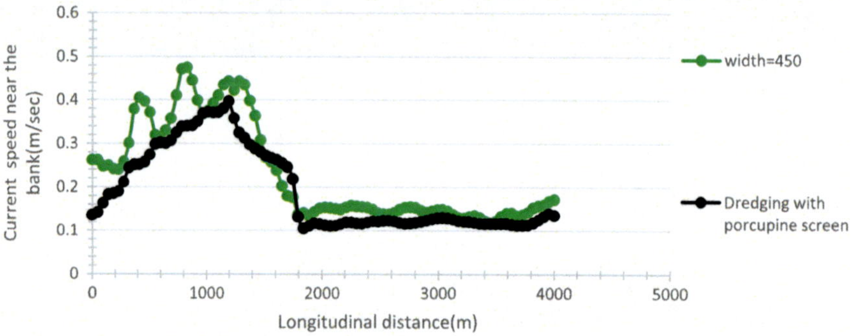

(b) Current speed near the bank with series of RCC porcupine screens

Fig. 4 **a** Proposed location for RCC porcupine screen installation. **b** Current speed near the bank with series of RCC porcupine screens

6 Conclusions

A 2D hydrodynamic modelling study is carried out by using BRAHMA2D model in the Brahmaputra River for a reach length of 35 km near Nimatighat to find out best possible measures for bank protection. Two possible alternatives are used in the simulation. In the first case, mid sand bar is dredged with three possible widths and computed velocity profiles are compared. In the second case, a series of RCC porcupines are installed at the upstream of the dredged channel. From the simulation, it is observed that the 450 m dredged channel considerably reduces the velocity near the bank; however, no significant changes is noticed in flow speed after the installation of the porcupine screens. It is necessary to carry out a morpho-hydrodynamic and cost–benefit analysis before implementing these methods in the field.

References

1. Church M, Ferguson RI (2015) Morphodynamics: Rivers beyond steady state. Water Resources Res, 51 (4):1883–1897 https://doi.org/10.1002/2014wr016862
2. Leopold LB, Maddock T (1953) The Hydraulic Geometry of Stream Channels and Some Physiographic Implications. Geological Survey Professional Paper 252
3. Liang D, Lin B, Falconer RA (2007) A boundary-fitted numerical model for flood routing with shock-capturing capability. J Hydrol 332:477–486
4. Anderson DA, Tannehill JD, Pletcher RH (1984) Computational fluid mechanicsand heat transfer. McGraw-Hill, New York
5. Saikia MD, Sarma AK (2006) Analysis for adopting logical channel section for 1D dam break analysis in natural channels. ARPN J Eng Appl Sci 1(2):46–54
6. Kalita HM (2016) A new total variation diminishing predictor corrector approach for two-dimensional shallow water flow. Water Resour Manage

Chapter 11
Flood Modeling in River System Using Gamma Memory

Agarwal Shivam[1](✉), Choudhury Parthsarathi[1], Roy Parthajit[1], and Debbarma Nilotpal[2]

[1] Civil Engineering Department, NIT Silchar, Assam, India
[2] Civil Engineering Department, NIT Agartala, Tripura, India

Abstract. In this study, applications can be seen regarding artificial neural network (ANN)-based models in predicting flow depth and flow rate for multiple bounding sections in river basin. Flow rate and depth are measured contemporaneous for Tar river basin section for the period of two months are forecasted by using a single ANN and as many ANNs as that of sections. Multiple-inputs multiple-outputs MIMO networks: MIMO-1 ANN and MIMO-2 ANN and multiple-inputs single-output MISO ANN are applied for obtaining flow forecasts for a number of sections in a river basin. This model is based on explicitly and implicitly learning storage properties during unsteady flow in river reaches. MIMO-1 and MIMO-2 ANN learn fractional and actual storage variation characteristics, respectively, during a period and obey continuity norms, while MISO ANN does not obey the continuity principle and learns an arbitrary storage variation to forecast concurrent flows in a river system. To test applicability of MIMO-1, MIMO-2 and MISO model forms, ANNs having memory and no memory are used to forecast concurrent flows rates and flow depths for multiple sections in Tar river basin, USA. Flow rates and flow depth values for four different sections in Tar river basin, USA, are forecasted using: multilayer perceptron (MLP), time delay neural networks (TDNN) and multiple gamma memory neural network (MGMNN) models. To develop the forecasting models that learn storage variation in the reach explicitly from the storage characteristic data of the data sets, viz. 'rate of storage change' and 'average flow depth' data for the river systems. Which are used along with flow rate/flow depth data in training the networks. Model performances are evaluated by using root mean squared error, coefficient of efficiency and the peak flow criteria.

Keywords: River systems · Gamma memory · Artificial neural network (ANN) · Time delay neural network (TDNN)

1 Introduction

It is known that many natural phenomena and climate variables like solar activity, temperature, sea level, icethickness, and many others are changing cyclically. Random factors

R. K. Bhattacharjya et al. (eds.), *Sustainable Water Resources Management*, Advances in Sustainability Science and Technology,
https://doi.org/10.1007/978-981-16-7535-5_11

are superimposed on the regularity of variability of these variables. Very urgent is the need for effective forecasting of changes in all cyclical, including river flow, variables. Typically, the prediction of such processes is based on the use of extrapolation of trends in the time series of dynamics. However, such a traditional approach does not allow the full use of the background of long and poorly structured time series characteristic of natural phenomena. In this regard, it is promising to use a new approach based on time delay neural network and gamma memory neural network modeling. However, issues of this approach are not currently developed. In this regard, this paper suggests applicability of MIMO-1, MIMO-2 and MISO model forms ANNs having memory and no memory to forecast concurrent flows rates and flow depths for multiple sections in TAR River Basin, USA. Also, the use of storage rate change variable which is an important aspect that needs to be incorporated in unsteady flow which is also taken into consideration.

Flood warning system protects people's lives and property and is an important means to reduce flood disasters. Development needs to be based on high-efficiency and high-precision flood forecasting models. Short intervals and fast arrival times of flood peaks make flood forecasting another important aspect. The difficulty is greatly increased. Therefore, in order for decision makers to have time to make flood warnings, flood evacuation and disaster relief basis of measures, efficient and highly accurate flood-level forecast models are much needed. However, the most important issue in flood forecasting is the key that lies in the temporal and spatial changes of storage forecast. Therefore, this study aims to quantify river flow estimates from single points on the ground, spatial and temporal variations of river flow, and gauge height to establish a storage-water level forecasting model through ANN to establish a storage water-level forecasting model and water level is included in the calculation of the hydraulic numerical model, adjust the system output in time to make. The residuals between the calculated and observed water levels and storage rate change are minimal. Also, the technique of gamma memory neural network can grasp the river flow in the watershed and spatial changes to enhance storage water-level forecasting models and river floods. The water calculation model's accuracy for floods provides the period of flow data, accurate and detailed flood information to issue flood alerts, flood evacuation strategy, and emergency response measures.

The main reasons for the difficulties in the research and hydrological forecasts are the complexity and uncertainty of the hydrological phenomenon itself, as well as the complicated internal nonlinear relationship [1]. To solve these problems, this paper establishes a neural network time series model and uses gamma memory to connect the neural network to the right values and optimizing various variables. This model is used to predict the two-hourly natural flow of the Tar river basin, North Carolina, from July 29, 2004, to October 1, 2004, and verify the test results with various statistical criteria like coefficient of correlation, etc. The study demonstrates that adhering to the continuity norms concurrent flows in multiple sections in a river basin can be forecasted by using MIMO-1 and MIMO-2 ANN models. The study indicates that MGMNN having adaptive memory depth is more suitable in cases where characteristics of the input signals are not known apriori. Having adjustable memory depth which is updated during training period, an MGMNN learns how to forecast and also learns subsequently how to make the forecast best. For TDNN and MLP, as the static memory depths are pre-selected, it may not match the characteristics of the unknown input signals leading to poor model performances [2]. The study extends applications of ANN models in forecasting multiple flows/flow depths in a river system; extends applicability of temporal ANN in modeling flood flow through river reaches and demonstrates the input–output data arrangement required for training an ANN for learning both storage and flow characteristics of the bounding sections in a river reach/system during unsteady flow such that continuity requirement is always obeyed [3].

2 Concurrent Flow Forecasting Approach

Concurrent Flow Predicting Models

During unsteady flow in river reaches, storage at the bounding sections dynamically changes over time. For river reach, storage S_t can be defined as a function of the concurrent flow rates as given

$$S_{(t)} = f_1\left(Q_t^u, Q_t^d, \psi\right) \tag{1}$$

where

$S_{(t)}$ = storage at time t

Q_t^u = flow rate at the upstream section at time t

Q_t^d = flow rate at the downstream section at time t

ψ = channel reach properties

Equation 1 depicts a system in which the upstream flow and downstream flow along with channel storage dynamically changing over a period of time depending upon the river basin characteristics [2].

Considering a small time period Δt, on the basis of Eq. 1, the following relationship can be drawn.

$$S_{(t+\Delta t)} = f_1\left(Q_{t+\Delta t}^u, Q_{t+\Delta t}^d, \psi\right) = \phi\left(S_{(t)}\right) = \phi' f_1\left(Q_t^u, Q_t^d, \psi\right) \tag{2}$$

where $Q_{t+\Delta t}^u, Q_{t+\Delta t}^d$ = upstream and downstream flow rate at time $(t + \Delta t)$.

Equation (2) shows that concurrent flow rates at the downstream and upstream stations having time interval Δt are related [4]. Based on the above functional relationship, an initial flow rate for the upstream and the downstream station given by αQ_t^u and βQ_t^u, respectively, as determined by the section characteristics, (φ) represented by α, β and reach properties(ψ), can be selected that produces no flow at one of the bounding sections after the time interval Δt. And, as given by the continuity norm, another set of initial flow given by $(1 - \alpha)Q_t^u$ and $(1 - \beta)Q_t^d$ for the upstream and downstream stations, respectively, can be obtained such that 'no flow' occurs at the other bounding section after the time interval, Δt. Using these two characteristic flow variations for a river reach in Eq. 2, the following functional forms can be obtained

$$Q_{t+\Delta t}^u = f_1\left(Q_t^u, Q_t^d, \psi, \varphi\right) \tag{3}$$

$$Q_{t+\Delta t}^d = f_2\left(Q_t^u, Q_t^d, \psi, \varphi\right) \tag{4}$$

In the case of the Muskingum model as per [2, 5], Eqs. 3 and 4 are obtained as given in Eq. 5 and Eq. 6, respectively.

$$Q^u_{t+\Delta t} = \frac{1}{-(1-c_1-c_3)}\left(c_1\alpha Q^u_t + c_3\beta Q^d_t\right) \tag{5}$$

$$Q^d_{t+\Delta t} = c_1(1-\alpha)Q^u_t + c_3(1-\beta)Q^d_t \tag{6}$$

Here, c_1, c_3 = Muskingum model parameters and α, β = upstream hydrograph evolution parameters and define the initial flow condition at the upstream and the downstream stations that produces no downstream flow after a time interval, Δt [5]. The characteristic flow variations at the upstream and the downstream stations represented by Eqs. 5 and 6 are $[\alpha Q^u_t \longrightarrow Q^u_{t+\Delta t}]$ & $[\beta Q^d_t \longrightarrow 0]$ and $[(1-\alpha)Q^u_t \longrightarrow 0]$ and $[(1-\beta)Q^d_t \longrightarrow Q^d_{t+\Delta t}]$, respectively. As given by the Muskingum model, Eqs. (5) and (6) establish the functional relationship existing between future flow at time $(t + \Delta t)$ at a section and the current flow rates at time t in the reach. It can be found that the fractional storage changes depicted by Eq. (5) and Eq. (6) are complimentary and sum to the total storage change in a river reach during the time interval, Δ [2]. Equations (5) and (6) give models that can be used to forecast u/s and d/s flow rates in a river reach, respectively. The models can be defined for a river reach estimating the model parameters which are $c_1, c_2, c_3, k, x, \alpha$ and β for a river reach when Muskingum model is considered [4].

The forecasting models developed for a river reach by learning characteristics of the fractional storage changes are even though MIMO ANNs but as they give forecasts for only one section, they may be referred as MIMO-1 ANNs. Merging two MIMO-1 networks depicted by Eqs. (5) and (6) a different MIMO ANN having two input and two output nodes can be obtained. Such networks when trained with inputs Q^u_t, Q^d_t, and desired outputs $Q^u_{t+\Delta t}$, $Q^d_{t+\Delta t}$ learn implicitly characteristics of the actual storage change during the period, Δt and may be termed as MIMO-2 ANNs. Further, as given by Eqs. (5) and (6), two-inputs one-output networks may be trained with Q^u_t, Q^d_t as inputs and $Q^u_{t+\Delta t}$, $Q^d_{t+\Delta t}$ as the desired output obtaining upstream/downstream flow forecasting model for a reach. These may be referred as multi-inputs single-output (MISO) form of ANNs.

Network Architecture

The ANN architecture for the models is observed using a supervised learning technique in which the series of flow data for a river system consisting of $N + 1$ number of flow depth and rate are split into three sets, training, calibration and testing. First 60% of the sequential data is used for training, to monitor the performance of the network during training next 20% of the sequential data is used for calibration and finally, the last 20% is used for measuring the model performances. ANNs being experience based learner from the data sets the representation of basin geomorphology through non-linear flow series and depth are considered for flood flow analysis. The sequential concurrent flow rate data at time t for $N + 1$ flow series of the river system is used as input for the networks like TDNN and MGMNN, but MLP-I being instantaneous mappers and having no memory, the concurrent flow rates data are randomized first and input–output sets are selected from the randomized data series to train the network.

Relationship Between Discharge and Depth of Flow

The relationship between the discharge $Q_{(t)}^{(*)}$ passing through a river section and the depth of flow $y_{(t)}^{(*)}$ at the sections is usually described by a power relation given as

$$Q_{(t)}^{(*)} = \omega\left(y_{(t)}^{(*)}\right)^{\eta} \tag{7}$$

It may be noted that with known flow characteristics as defined by the parameters ω and η, the discharge passing through a section can be estimated on the basis of the recorded flow depth [4]. The estimate for the flow rate written in terms of flow depth and section characteristics may be used in the flow models given by Eqs. (2), (5) and (6) to reformulate the models using depth variable. That is, by using the relation (7) in conjunction with Eqs. (5) and (6), the 'end of the period' flow depth for the upstream and downstream stations can be written as a function [5, 6] of the current flow depths at the upstream and downstream station as given is Eqs. (8) and (9)

$$y_{t+\Delta t}^{u} = \left\{\frac{1}{-(1-c_1-c_2)\omega_u}\left(c_1\alpha\omega_u\left(y_t^u\right)^{\eta^u}\right) + c_3\beta\omega_d\left(y_t^d\right)^{\eta^d}\right\}^{1/\eta^u} \tag{8}$$

$$y_{t+\Delta t}^{d} = \left\{\frac{1}{\omega_d}\left(c_1(1-\alpha)\omega_u\left(y_t^u\right)^{\eta^u}\right) + c_3(1-\beta)\omega_d\left(y_t^d\right)^{\eta^d}\right\}^{1/\eta^d} \tag{9}$$

Comparing Eqs. (5) and (6) with Eqs. (8) and (9), it can be found that the degree of nonlinearity for flow depth forecasting models is higher involving increased number of model parameters. In that case, optimal estimation of parameters using traditional methods may be difficult resulting in erroneous forecasts. However, as evident from the input–output data structure depicted by Eqs. (8) and (9), arrangement of upstream and downstream flow depths can be made to train an ANN for learning the functional nonlinearity among flow variables and implicitly learning the storage change characteristics from the flow depth data sets.

In the case of a river system having N upstream flows, Eqs. (5) and (6) can be written using equivalent inflow as [2, 4, 6]

$$Q_{t+\Delta t}^{u,e,r} = \frac{1}{-(1-c_1-c_3)}\left(c_1\alpha Q_t^{u,e,r} + c_3\beta Q_t^d\right) \tag{10a}$$

$$Q_{t+\Delta t}^{d} = c_1(1-\alpha)Q_t^{u,e,r} + c_3(1-\beta)Q_t^d \tag{10b}$$

$$Q_t^{u,e,r} = \sum_{p=1}^{N}\sigma^{p,r}Q_t^p \tag{10c}$$

Here, $Q_t^{u,e,r}$ = the equivalent inflow at a point r in the basin for N upstream flows measured at different locations and $\sigma^{p,r}$ = the shift factor associated with the transfer of flow from p to r. Considering the above fractional storage variations for a river system each upstream flow and the common downstream flow at time, $(t + \Delta t)$ can be written explicitly as a function of flow rates at time t for all sections in the system as given in Eqs. (11a) and (11b) [2, 6]

$$Q^{u,p}_{t+\Delta t} = f^n\left(Q^{u,1}_t, Q^{u,2}_t, Q^{u,3}_t, \ldots, Q^{u,N}_t; Q^d_t, \psi, \phi\right); \forall p; p = 1, 2, 3, \ldots, N \quad (11a)$$

$$Q^d_{t+\Delta t} = g\left(Q^{u,1}_t, Q^{u,2}_t, Q^{u,3}_t, \ldots, Q^{u,p}_t, \ldots, Q^{u,N}_t; Q^d_t, \psi, \phi\right) \quad (11b)$$

Again applying flow depth variables for the bounding sections in a river system equations, (11a) and (11b) can be obtained as given Eqs. (11c) and (11d) [2, 5]

$$y^{u,p}_{t+\Delta t} = g^n\left(y^{u,1}_t, y^{u,2}_t, y^{u,3}_t, \ldots, y^{u,N}_t; y^d_t, \psi', \phi'\right); \forall p; p = 1, 2, 3, \ldots, N \quad (11c)$$

$$y^d_{t+\Delta t} = G\left(y^{u,1}_t, y^{u,2}_t, y^{u,3}_t, \ldots, y^{u,p}_t, \ldots, y^{u,N}_t; y^d_t, \psi', \phi'\right) \quad (11d)$$

Like a river reach, there are $(N + 1)$ number of MIMO-1 and one MIMO-2 ANN models for a river system for forecasting flow depth and flow rate variables. Similarly, there can be $(N + 1)$ a number of MISO ANN models that by learning arbitrary storage variations forecast flow rates/flow depths for $(N + 1)$ sections in a river system. The MIMO-1 and MIMO-2 models when trained using input–output flow rate/depth data sets separated by a time duration Δt learn the storage evolution characteristics for the reach implicitly. To make MIMO-1 and MIMO-2 ANNs learn storage characteristics explicitly along with learning, the flow characteristics for a reach the ANNs may be trained with data sets that reflect storage characteristic along with the flow data sets.

3 Results and Discussions

Performances such as root mean square error, normalized mean squared error, coefficient of correlation, mean absolute error; minimum and maximum absolute error are represented in Table 3 for the MIMO-2 model form. Performance results and forecasted flow rates in predicting the series having two hour of lead time are represented in Fig. 1. Results depict that flow rate and storage rate change forecasted using MGMNN are somewhat matching with the observed flow rate of the sections. From the results, it can be said that from a given series of data, MIMO-2 models form can very well learn from the training data sets. MIMO-1 model on the other hand provides accuracy in matching zero flow rate condition Table 1 which helps in accounting fraction variation of storage while following the principle of continuity. Figure 3 depicts the study area of Tar river basin, USA, in which Enfield, Hilliardston, Rockymount are considered as inflow stations while Greenvile and Tarboro are considered downstream stations. Table 2 depicts the peak flow criteria in MIMO-1 model forms for various forecasting stations using MLP, TDNN and GMNN architectures. Peak flow is an important statistical parameter which helps in formulating the occurrence of flash flood. It also predicts about the duration and period of maximum discharge that could possibly lead to floods in a short notice. Hence, calculation of PFC is an important aspect in river flow modeling. Values obtained in Table 2 clearly state that there is less to moderate chances of flow variation that prevailed during August to October in 2004.

$$PFC = \frac{\left(\sum_{t=1}^{T_p}\left(Q_t^{o,r} - Q_t^{f,r}\right)\left(Q_t^{o,r})^2\right)^{\frac{1}{4}}}{\left(\sum_{t=1}^{T_p}(Q_t^{o,r})^2\right)^{\frac{1}{2}}} \tag{12}$$

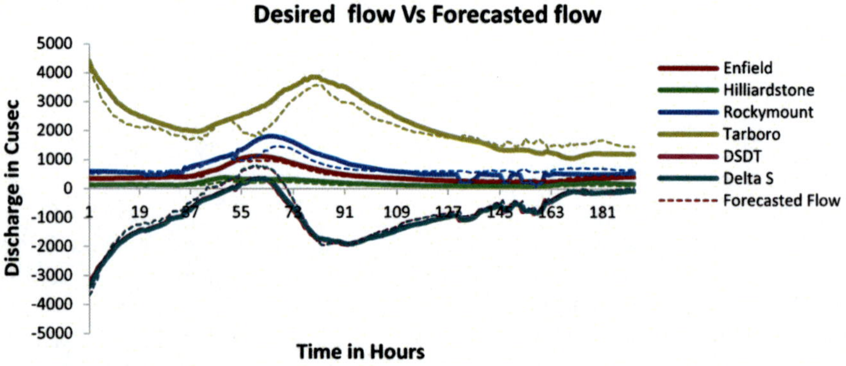

Fig. 1 Desired and 2-h ahead predicted flows for various sections in Tar River basin obtained by using MIMO-2 MGMNN

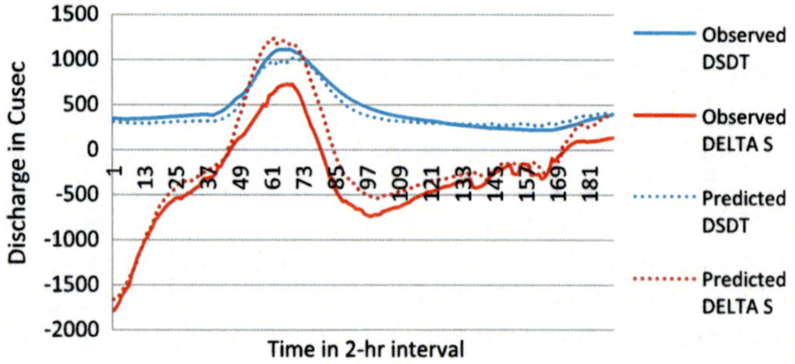

Fig. 2 Observed and one time step ahead forecasted instantaneous storage rate and Delta S flow at Tarboro (starting from 00:00 h of 29th July) using MIMO-1 MGNN

4 Conclusions

Daily-based hourly scale river forecasting has significant seasonality and trend. This paper uses a neural network with gamma memory. The inter-sequence model establishes a forecasting model. The daily hourly flow of the Tar-Pamlico River Basin in North Carolina is forecasted and tested. The prediction results of the multi-layer perceptron (MLP) neural network and gamma neural network are compared, and the following

Table 1 Performances in terms of RMSE of MIMO-1 MLP, GMNN and TDNN for matching zero flow rate

Model	Forecast section	Bounding sections			
MLP-1		TAR	RMT	HLD	ELD
	TAR	–	0.006	0.007	0.012
	RMT	0.006	–	0.010	0.008
	HLD	0.013	0.004	–	0.016
	ELD	0.016	0.010	0.000	–
MGMNN	TAR	–	0.065	0.032	0.037
	RMT	0.003	–	0.019	0.009
	HLD	0.022	0.020	–	0.006
	ELD	0.009	0.013	0.025	-
TDNN	TAR	–	0.009	0.012	0.066
	RMT	0.016	–	0.027	0.017
	HLD	0.018	0.008	–	0.008
	ELD	0.006	0.009	0.012	–

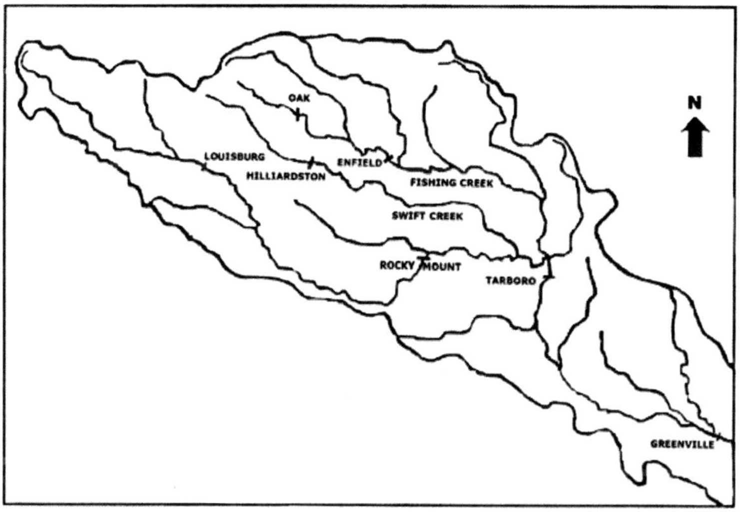

Fig. 3 Map of all the gauging stations in Tar River Basin of study area

Table 2 Model performances in terms of PFC for forecasting flow rates by using MIMO-1 form of ANN models in Tar basin

Forecasting stations	Models	DSDT (storage rate change)	DELTA storage
Enfiled	Gamma	0.0913	0.1366
	MLP	0.1061	0.0916
	TDNN	0.1058	0.1650
Rocky mount	TDNN	0.1254	0.1556
	Gamma	0.1292	0.1556
	MLP	0.0898	0.1335
Tarboro	Gamma	0.1080	0.0864
	MLP	0.0515	0.0517
	TDNN	0.1087	0.1090
Hillardstone	Gamma	0.1669	0.1304
	MLP	0.1232	0.1021
	TDNN	0.0987	0.0950

Table 3 Performances of MIMO-2 Models based on various statistical criterions

Performance	Enfield	Hilliardstone	Rockymount	Tarboro	d*s*d*t*	Delta *S*
RMSE	63.38	66.08	217.12	393.99	211.73	209.41
NMSE	0.06	0.63	0.28	0.20	0.07	0.07
MAE	52.83	53.84	172.95	341.69	171.16	169.29
Min Abs Error	0.06	1.53	1.50	0.66	1.54	0.79
Max Abs Error	167.18	146.56	572.23	983.92	543.57	503.69
R	0.98	0.96	0.92	0.94	0.98	0.98

conclusions are drawn: (1) The prediction accuracy is significantly higher than the traditional MLP network which is why MLP with sliding window approach is equivalent to time delay neural network TDNN; (2) the gamma memory neural network is more accurate for the flow rate and gauge height than the simple ANN network using static MLP. To predict the trend of the time series of the stream, (3) the time delay neural network with the gamma operator can obviously overcome the traditional multi-layer neural network, and gamma network is easy to fall into a local minimum defect and has better stability. In general, this gamma algorithm neural network optimized in time series model can well predict hourly flow.

References

1. Aboutalebi M, Haddad OB, Loáiciga HA (2016) Application of the SVR-NSGAII to hydrograph routing in open channels. J Irrig Drain Eng 142(3):04015061
2. Choudhury P, Roy P (2015) Forecasting concurrent flows in a river system using ANNs. J Hydrol Eng 20(8):06014012
3. Sil BS, Choudhury P (2016) Muskingum equation based downstream sediment flow simulation models for a river system. Int J Sedim Res 31(2):139–148
4. Choudhury P, Ullah N (2014) Downstream flow top width prediction in a river system. Water SA 40(3):481–490, 982–994
5. Choudhury P (2007) Multiple inflows Muskingum routing model. J Hydrol Eng 12(5):473–481
6. Choudhury P, Sankarasubramanian A (2009) River flood forecasting using complementary Muskingum rating equations. J Hydrol Eng 14(7):745–751

Chapter 12
2D Hydrodynamic Model for Evaluating Impact of Possible Riverfront Activities in an Urbanized Bank of Brahmaputra River

Gaurav Talukdar(✉), Anupal Baruah, and Arup Kumar Sarma

Department of Civil Engineering, Indian Institute of Technology, Guwahati 781039, India
{gauravt,anupalbaruah,aks}@iitg.ac.in

Abstract. Hydrodynamic modelling of channels can be used as a decision-making tool for different works. For large channels, implementation of the numerical models for estimation of flow and the other parameters are cost effective and less tedious than the physical models. Model study can be used in the design of hydraulic structures, estimation of scouring near the flow obstacles and in the flood inundation mapping. In the present work, a hydrodynamic model study is carried out for 1.00 km reach length of Brahmaputra River near Guwahati city of Assam. The main objective of the study is to understand changes in the flow scenario under two different discharges due to implementation of some riverfront activities on the urbanized bank of the river Brahmaputra. The range of scour depth near the bank in the modelled area is also estimated and presented. The impact of the considered riverfront activities on the flow scenario are evaluated in terms of variation in water level and velocity vector using a 2D hydrodynamic model. The variation in flow scenario due to such intervention has shown some minor variation in the flow.

Keywords: Riverfront · Brahmaputra river · Mathematical modelling · Scour depth · Velocity

1 Introduction

Socioeconomic upliftment due to rapid urbanization in the metro areas is a prime concern in today's era. The infrastructure development in the cities helps in the growing economy of the individuals as well as enhancing the lifestyle of the citizens. In the north-eastern part of the country, the Brahmaputra River provides immense opportunity in the development of the tourism as well as in some other sectors. The riverfront protection or development in this dynamic river is a critical task because of its unpredicted nature. Near the city of Guwahati, during monsoon, generally the river carries more than 40,000 cumecs of water with sediments, whereas during lean period, the flow recedes to near about 3000 cumecs. Actual variation in flow however ranges from nearly 2000 cumec to

R. K. Bhattacharjya et al. (eds.), *Sustainable Water Resources Management*,
Advances in Sustainability Science and Technology,
https://doi.org/10.1007/978-981-16-7535-5_12

70,000 cumec in this reach. A detail study is needed to be carried out before implementing any near bank infrastructure developmental work at the waterfront of Brahmaputra River. Physical modelling of a braided river is not quite cost effective and is tedious in nature. Under this circumstance, numerical modelling is a better alternative to the former with some amount of uncertainty. Out of different hydrodynamic modelling tools, MIKE21C model is used for the simulation in this study. MIKE21C is a 2D software developed by the Danish Hydraulics Institute, Denmark, used for simulating free-surface flows, water quality, sediment transport and waves in rivers, lakes, estuaries, bays, coastal seas and other water bodies. It solves the 2D depth averaged unsteady free surface flow equations in curvilinear coordinates by finite difference implicit method. Hassan and Dibike [2] used MIKE21C to simulate the confluence of the rivers Ganges and Yamuna, situated in Bangladesh, and showed its potential in river dredging and navigation studies. Hye and Jahan [3] used MIKE21C to simulate Meghna River in Bangladesh and investigate the response to flow and sediment regime changes due to construction of a road bridge upstream of an existing railway bridge. Beck and Basson [1] used MIKE21C to determine the effectiveness of flushing of sediments during breaching, by investigating the breaching process at different water levels for the Klein river estuary in South Africa.

In the present study, a 2D hydrodynamic modelling is done at a reach length of 1.00 km at Brahmaputra River near Guwahati. Simulation is carried out for two scenarios to study the possibility of changes in the flow characteristics near the bank due to possible development/protection activities taken up at the riverfront. A critical discussion is made based on the observed results. Figure 1 shows the satellite images of the modelled reach in different years.

2 Study Area and Model Description

Brahmaputra River is one of the largest rivers in the world carrying huge amounts of sediments. The river crosses the Indo-China Border near Pasighat, entering Assam and flows westward for about 700 km downstream towards Dhubri leaving Assam and entering Bangladesh. For the present study, a river reach of around 1.00 km on the main stem Brahmaputra has been considered (Fig. 2). The area lies between 91° 44′ 28.51″ E, 26° 12′ 05.91″ N and 91° 42′ 04.45″ E, 26° 10′ 15.64″ N in the Guwahati region. The bed slope in this portion is around 0.1 m/km. The temperature varies from 10 °C in the winters to 38 °C in the summers.

The advancement of the study in understanding the changes within the river and its effects has been possible through model study. Hydrodynamic models solve complex equations using numerical methods with the advantage of calculating various flow parameters. MIKE 21C has been used to carry out the simulations in different scenarios in the present study.

MIKE 21C simulates the 2D flow in curvilinear grid and studies the sediment transport and morphological processes in rivers. It is a 2D depth averaged flow model based on the Saint Venant equation (continuity and conservation of momentum). The equations solved are:

$$\frac{\delta p}{\delta t}+\frac{\delta}{\delta s}\left(\frac{p^2}{h}\right)+\frac{\delta}{\delta n}\left(\frac{pq}{h}\right)+2\frac{pq}{hR_n}+\left(\frac{p^2-q^2}{hR_s}\right)+gh\frac{\delta H}{\delta s}+\frac{g}{C^2}\frac{p\sqrt{p^2+q^2}}{h^2}=\text{RHS}$$

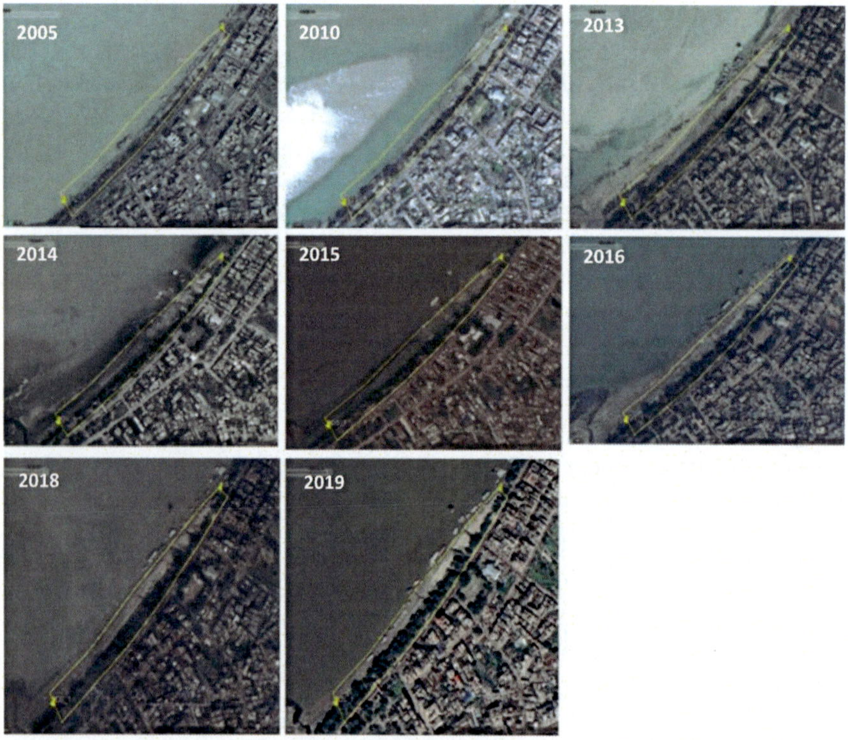

Fig. 1 Historical satellite imageries depicting the site for riverfront development

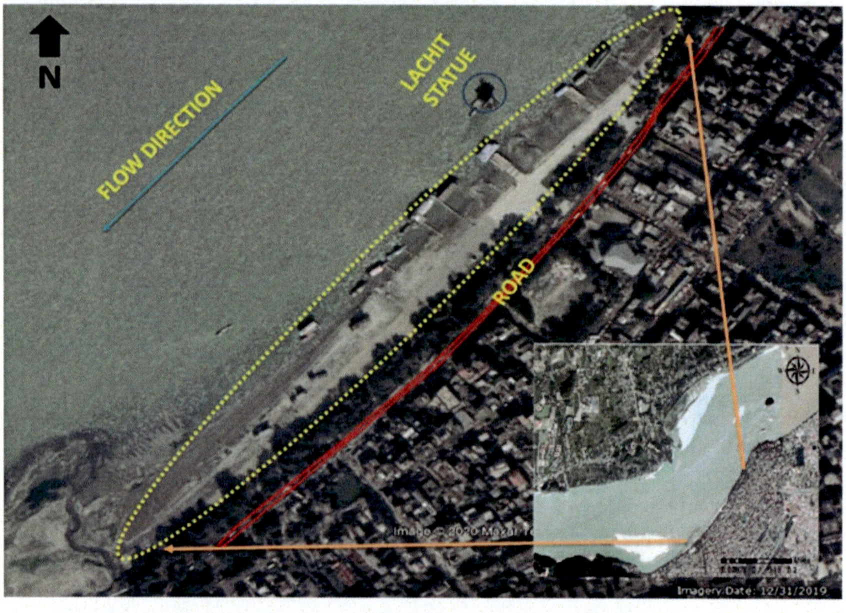

Fig. 2 Study area of the Brahmaputra River, Assam

$$\frac{\delta q}{\delta t} + \frac{\delta}{\delta n}\left(\frac{q^2}{h}\right) + \frac{\delta}{\delta s}\left(\frac{pq}{h}\right) + 2\frac{pq}{hR_s} + \left(\frac{q^2 - p^2}{hR_n}\right) + gh\frac{\delta H}{\delta n} + \frac{g}{C^2}\frac{q\sqrt{p^2+q^2}}{h^2} = \text{RHS}$$

$$\frac{\delta H}{\delta t} + \frac{\delta p}{\delta s} + \frac{\delta q}{\delta n} - \frac{q}{R_s} + \frac{p}{R_n} = 0$$

where *s*,*n*—coordinates of curvilinear coordinate system, *p*,*q*—mass fluxes in *s* and *n* direction, *H*—water level, *h*—water depth, *g*—gravitational acceleration, *C*—Chezy Roughness coefficient, R_s, R_d—radii of curvature of *s* and *n* lines, RHS—Reynold stresses.

3 Methodology and Applications

An important stage in simulation of the hydrodynamic model is the model set up and grid generation. A suitable curvilinear grid is developed over the study area whose coordinates are derived from the satellite imageries (Fig. 3). A dense grid provides accurate simulations of the water flow. For the model set up, a bathymetric file is required as an input over which the simulation run is dependent. The bathymetric survey was conducted along the study reach, and the simulation was performed with the necessary boundary conditions. The model calibrated and validated for this portion of the river reach under a different project was run at two different discharges given as an input in the upstream boundary condition. The observed water level at the downstream boundary for the corresponding discharges was used as the downstream boundary condition.

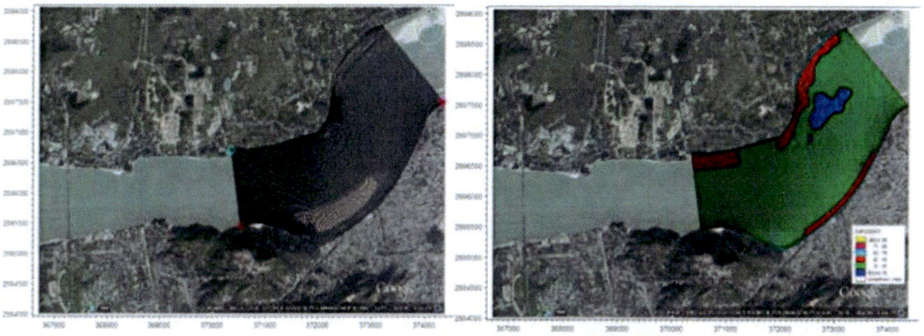

Fig. 3 Curvilinear grid and Bathymetry of the study area

The highest discharge recorded in the Brahmaputra River in Pandu (near Guwahati) is nearly 70,000 cumec and lowest recorded discharge is 1800 cumec (source: waterresources.assam.gov.in). The simulation was, however, performed for two different discharges: (a) considering high-flow scenario commensurate with the bank full discharge, which is about 50,000 cumec for the study area and (b) for a discharge of 30,000 cumec, at which the river water touches the bank with significant depth. Two scenarios were taken into consideration under which the entire study was performed. Scenario 1 represents the present condition of the river, wherein the area marked (Fig. 2) is studied for changing flow alteration near the bank. Scenario 2 represents a condition where the bank (encircled portion) is elevated to the road level and the flow conditions are checked. This study primarily focusses on evaluating changes in the present flow condition if the bank is raised in this relatively stable portion (marked in the Fig. 2) to develop the same for recreation and other social activities. This would attract investments, provide employment opportunities, tourism and add economic growth of the city.

4 Results and Discussion

It is observed from the historical images that the morphology of the modelled portion has not changed much for several years. Therefore, scope of utilizing this portion for various activities can be explored. The simulated water depth under different discharges are shown in Figs. 4 and 5 under the existing scenario and road elevated scenario. Figures 6 and 7 represent the simulated velocity vectors under the road elevated scenario and present scenario. An inundation in the modelled area is observed under high discharges for present condition, and therefore, raising of this portion is necessary to utilize that portion for some activities. Figures 8 and 9 show the scour depth at the u/s and d/s of the obstruction present in the reach for the present and proposed scenario. Under the proposed new scenario, no inundation is noticed in the modeled area; however, some variation in the velocity has been observed (Fig. 10).

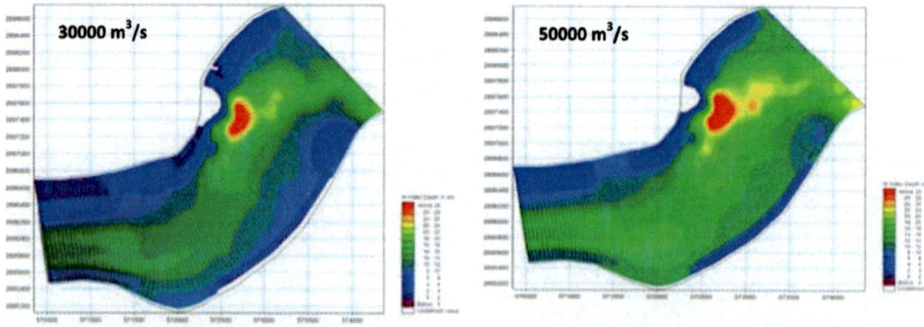

Fig. 4 Simulated water depth under varying discharges under the existing scenario

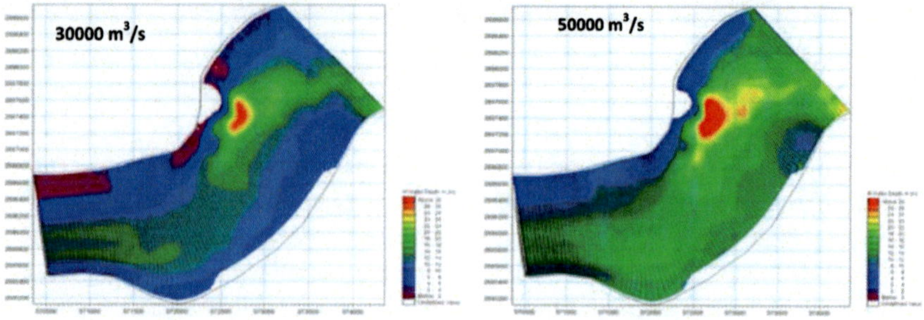

Fig. 5 Simulated water depth under varying discharges under the road-elevated scenario

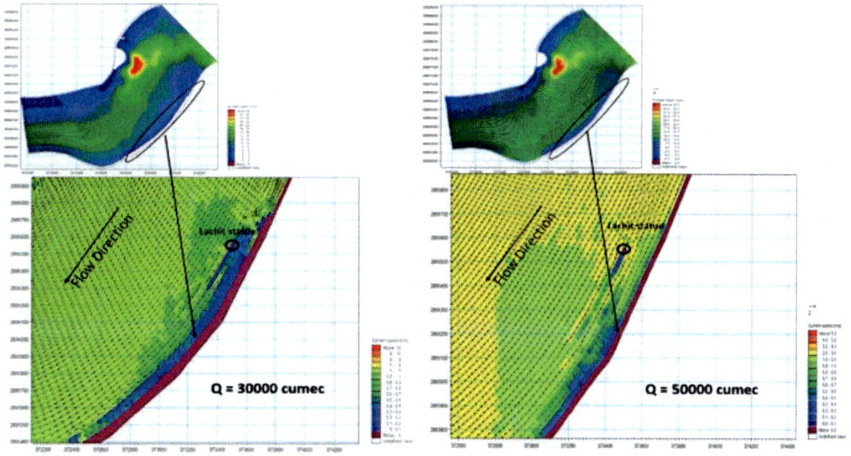

Fig. 6 Simulated velocity vectors under varying discharges under the present scenario

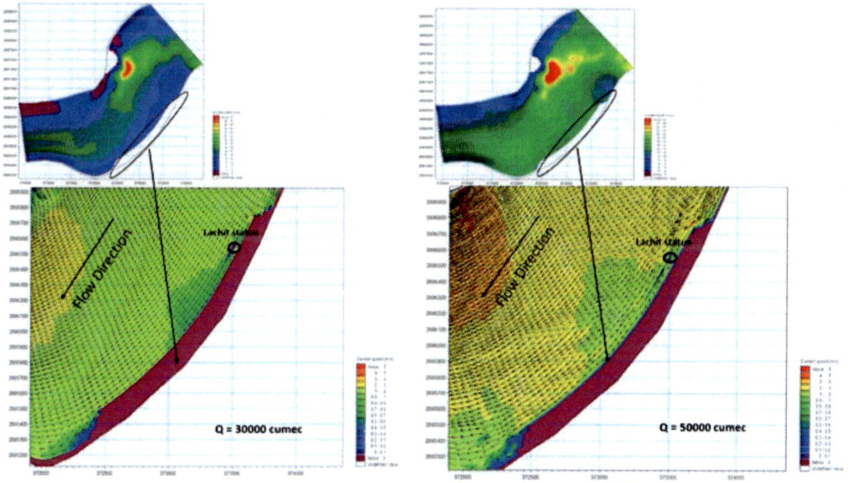

Fig. 7 Simulated velocity vectors under varying discharges under the road-elevated scenario

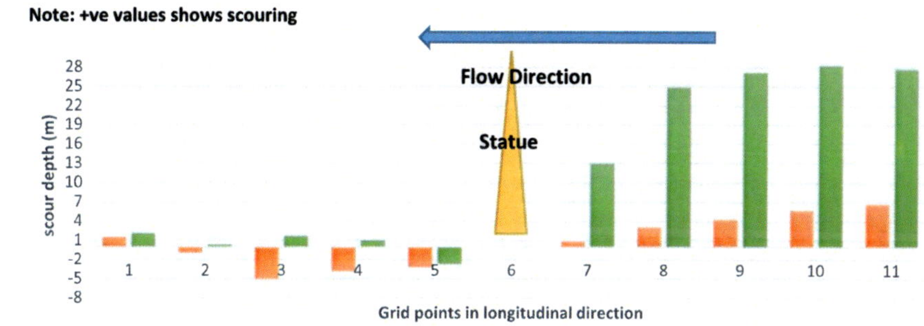

Fig. 8 Scour depth at u/s and d/s of the statue at present scenario **a** $Q = 30{,}000$ cumec, **b** $Q = 50{,}000$ cumec

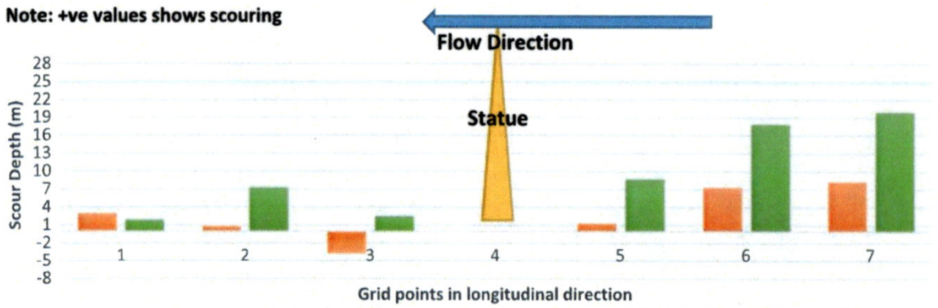

Fig. 9 Scour depth at u/s and d/s of the statue at proposed scenario **a** $Q = 30{,}000$ cumec, **b** $Q = 50{,}000$ cumec

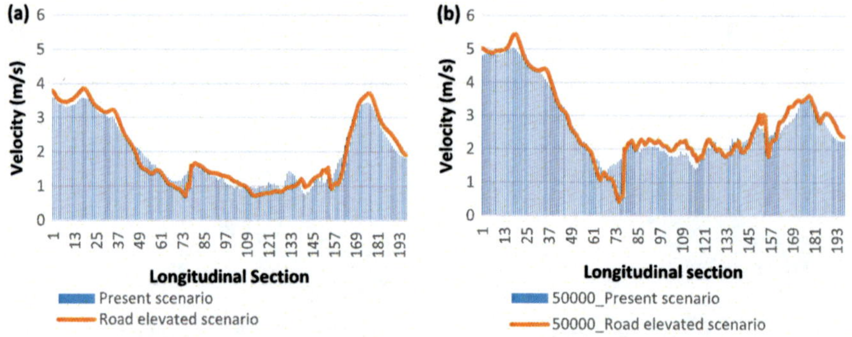

Fig. 10 Velocity plots of the present and proposed scenario at **a** 30,000 cumec and **b** 50,000 cumec

5 Conclusion

The geoinformatic study has shown that the stable area, identified from the historical planform study, can be protected from flooding and can be utilized for recreational and other economic activities to attract tourists. As the river is quite wide as compared to this narrow strip considered for raising, no significant changes in near bank velocity is noticed. The bank is not inundated at discharges of 30,000 and 50,000 cumecs under the proposed scenario. Of course, with higher discharges, flow will overtop. However, raising of the narrow stable strip to the present road level will prevent flooding of this area for a discharge up to 50,000 cumec. Maximum scour depth under the present scenario is observed within the range of 10–20 m and in the proposed scenario, scour depth is observed within the range of 7–15 m at the upstream. As the scour depth is high, necessary protection measures will have to be adopted for ensuring stability of the raised portion.

References

1. Beck JS, Basson GR (2008) Klein River estuary (South Africa): 2D numerical modelling of estuary breaching. Water S A 34(1):33–38
2. Hassan KI, Dibike YB (2000) Two-dimensional morphological modeling at the confluence of the Ganges and the Jamuna rivers for dredging and navigation study. In: Proc. 4th int. conf. hydroinf., Iowa, USA
3. Hye JMA, Jahan S (2000) Modeling response to flow and sediment regime changes due to construction of a bridge upstream of an existing bridge. In: Proc. int. cong. modeling and simulation, Canberra, Australia

Check for updates

Chapter 13
Optimization of Sukla Irrigation Canal

Sultana Hadia Rahman(✉) and Bibhash Sarma

Department of Civil Engineering, Assam Engineering College, Guwahati, Assam 781013, India

Abstract. Canals are the most important and major element for delivering water in an irrigation system. A substantial amount of investment is required depending on its length and cross section. So, cross-sectional design of the canal should be carried out on an optimization basis. An economic canal section is the one which requires the smallest amount of total construction cost. The total cost of construction of the canal includes excavation cost and surface lining cost, apart from labour and maintenance. So, open channel design is specially needed for determination of optimum dimensions of the canal and to construct the canal section with minimum cost of construction to convey the specified discharge. A nonlinear optimization model considering data of Sukla Irrigation Project is prepared. The objective function for the model is the total minimum cost of construction per unit length. The model is then solved in Microsoft Excel solver platform considering various constraints to obtain the minimum cost of construction per unit length of the channel and minimum dimensions of cross sections.

Keywords: Canal section optimization · Sukla irrigation project · Microsoft excel solver

1 Introduction

Canals are the most important element of an irrigation system. These are used to convey, distribute and apply water to the land. A substantial amount of investment is required depending on its length and cross section. Proper designing and planning, which result in an optimal cost, therefore, are of greatest importance in determining the economy of the project. An economic canal section is the one that needs the smallest amount of total construction cost which includes excavation cost and surface lining cost, apart from labour and maintenance. An open channel section having minimum area or least wetted perimeter for a given rate of flow, roughness coefficient and longitudinal slope is considered as the most economic channel section as it involves least amount of earth work and least lining surface. So, the design of open channel is required to determine the optimum dimensions of the canal and to construct the canal section with optimum cost to convey the specified discharge. A canal may be a rigid boundary (lined) canal and a mobile boundary (unlined) canal. Rectangular-lined canal is considered for designing as the lined canal permits higher velocity and the cross-sectional area required by lined

R. K. Bhattacharjya et al. (eds.), *Sustainable Water Resources Management*,
Advances in Sustainability Science and Technology,
https://doi.org/10.1007/978-981-16-7535-5_13

canal is less for a specific discharge. Moreover, rectangular canal section will further decrease the land area required by the canal. Again, the maintenance cost for lined canal is very less compared to unlined canals. The various parameters required to consider for designing are: (1) the roughness coefficient which is dependent on the material of lining; (2) the minimum permissible velocity to avoid deposition of silt and debris; (3) the maximum permissible velocity to avoid erosion in the channel surface; (4) bed slope; (5) hydraulic efficiency of the channel section; and (6) Froude number to maintain subcritical velocity.

Chow [3] listed various properties for most efficient hydraulic channel section. For a rectangular channel, if the depth is half of the width of the channel, it is hydraulically most efficient. Swamee et al. [7] developed a model for comprehensive design of minimum cost irrigation canal section. In their model, they considered the material forming canal section and minimum permissible velocity to avoid silting, limiting velocity to avoid erosion, hydraulic efficiency of channel section and the topography of the channel route to fix the bed slope. Bhattacharjya [1] developed an optimization model for optimal design of open-channel section incorporating critical flow condition. He developed a methodology to determine the cost-effective channel section considering water depth away from the critical section of the channel to avoid large fluctuation in water depth in the channel. Bhattacharjya and Satish [2] developed a new methodology to design a stable and optimal channel section by using hybrid optimization techniques. Hameed [4] used direct search optimization method by using MATLAB to solve the resulting channel optimization models for a specific flow rate, roughness coefficient and longitudinal slope and used his model in round bottom triangular channels.

In the present study, a nonlinear optimization model considering data of Sukla Irrigation Project is prepared. The objective function for the model is the total minimum cost of construction per unit length. The model is then solved in Microsoft Excel solver platform considering various constraints to obtain the minimum cost of construction per unit length of the channel and minimum dimensions of cross sections.

2 An Overview of Sukla Irrigation Project

Sukla Irrigation Project is located in Baksa district of Assam (Fig. 1) and one among the districts of Bodoland Territorial Area District. The headwork of the project has been created on a diversion channel of Sukla river close to village Naokata near Goreswar at latitude: 26°38′20″N and longitude: 91°44′15″E. The unlined canal system of the Sukla Irrigation Project contains two main canals D_1 and D_2 directly embark from the regulator. Main canal D_1 irrigates areas within the west of Baihata Chariali-Naokata-Goreswar PWD road, and the opposite distributary D_2 provides water to the east of the PWD road. The main canal D_1 has 12 branch canals, and the main canal D_2 has a total of 5 branch canals. There are a total of 4 distributaries in the branch canals of D_2. The schematic layout of the canal system is shown in Fig. 2.

The design discharge of the canal system is 24.42 m^3/s which is 75% dependable discharge calculated at Sukla headwork. The detail length of canal system, its designation and command area under each canal and required discharge by each canal of Sukla Irrigation Project is provided in Table 1.

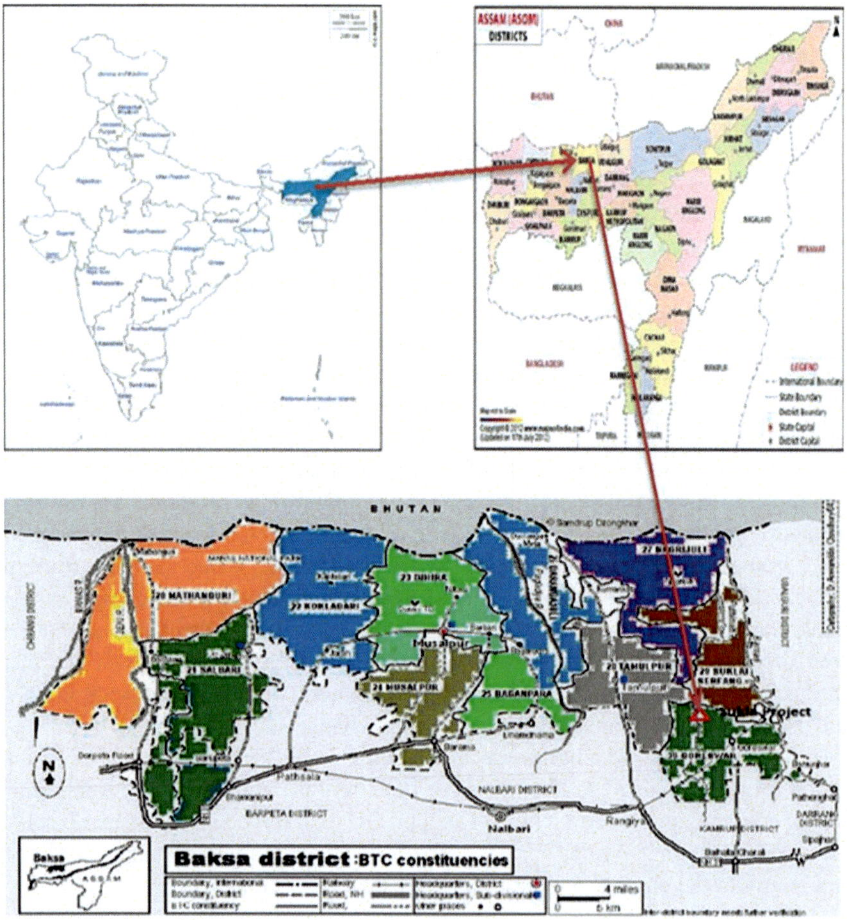

Fig. 1 Location map of Sukla Irrigation Project

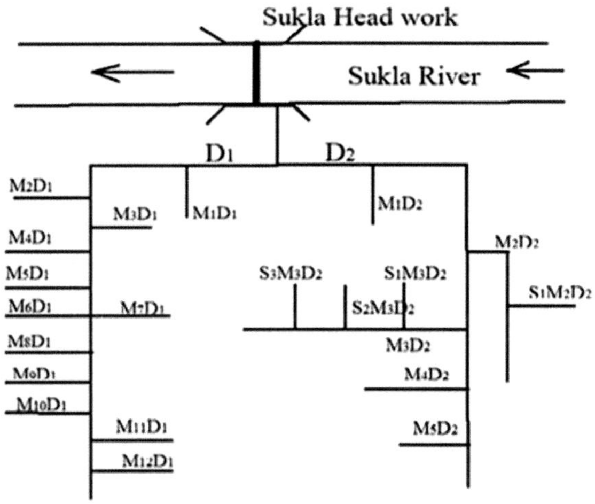

Fig. 2 Schematic layout of Sukla Irrigation Project

3 Materials and Methods

Consider a rectangular channel section as shown in Fig. 3, the Manning's roughness coefficient, flow depth, width of the channel, thickness of canal lining and free board of the channel are *n*, *Y*, *B*, *t* and *f*. Cement concrete is used as a lining material; hence, roughness coefficient for cement concrete is used.

Let A_f and P_f be the cross-sectional area and wetted perimeter of the channel with free board *f*.

$$\text{Area of the canal } A_f = B \times (Y + f)$$

$$\text{Perimeter of the canal } P_f = B + 2\,(Y + f)$$

Similarly, *A*, *P*, *R*, *T*, *D* and Fr are the area, wetted perimeter, hydraulic radius, top width, hydraulic depth and Froude's number of the channel without freeboard. These parameters can be written as

$$\text{Wetted area of the canal } A = B \times Y$$

$$\text{Wetted perimeter of the canal } P = B + 2Y$$

$$\text{Top width of canal } T = B$$

$$\text{Hydraulic depth } D = A/T$$

$$\text{Hydraulic radius of canal } R = A/P$$

Table 1 Length, command area and discharge required by the canal system of Sukla IP

Name of the canal	Designation	Length (m0	Bifurcation point at chainage (m)	Command area *A* (Ha)	Fraction of command area ($A/\sum A$)	Discharge required by the canal $A/\sum A \times 24.4$ (m^3/s)
Main	D_1	28,800		1920	0.167	4.077
Branch	M_1D_1	5974	4042.00 of D_1	200	0.017	0.425
	M_2D_1	2530	8107.68 of D_1	645	0.056	1.370
	M_3D_1	4980	9418.32 of D_1	234	0.020	0.497
	M_4D_1	1219	10,271.76 of D_1	176	0.015	0.374
	M_5D_1	2225	10,911.84 of D_1	232	0.020	0.493
	M_6D_1	6839.71	14,295.12 of D_1	686	0.060	1.457
	M_7D_1	1584.96	14,295.12 of D_1	48	0.004	0.102
	M_8D_1	3780	19,720.56 of D_1	659	0.057	1.399
	M_9D_1	2316	20,878.80 of D_1	262	0.023	0.556
	$M_{10}D_1$	2621	22,920.96 of D_1	234	0.020	0.497
	$M_{11}D_2$	914	23,032.20 of D_1	90	0.008	0.191
	$M_{12}D_1$	1524	24,932.64 of D_1	128	0.011	0.272
Main	D_2	31,700		2120	0.184	4.502
Branch	M_1D_2	5090	7943.00 of D_2	730	0.063	1.550
	M_2D_2	13,716	12,954.00 of D_2	1280	0.111	2.718
	M_3D_2	3658	19,933.92 of D_2	440	0.038	0.934

(*continued*)

Table 1 (*continued*)

Name of the canal	Designation	Length (m0	Bifurcation point at chainage (m)	Command area *A* (Ha)	Fraction of command area ($A/\sum A$)	Discharge required by the canal $A/\sum A \times 24.4$ (m^3/s)
	M_4D_2	3870.96	26,365.20 of D_2	265	0.023	0.563
	M_5D_2	1402	27,096.72 of D_2	215	0.019	0.457
Distributary	$S_1M_2D_2$	1798	6492.24 of M_2D_2	486	0.042	1.032
	$S_1M_3D_2$	1610	201.17 of M_3D_2	110	0.010	0.234
	$S_2M_3D_2$	1829	1188.72 of M_3D_2	194	0.017	0.412
	$S_3M_3D_2$	1524	2042.16 of M_3D_2	146	0.013	0.310
Total =				11,500	1	

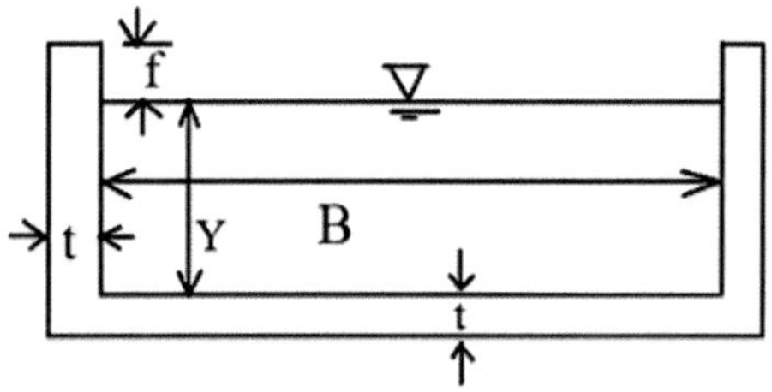

Fig. 3 Rectangular channel cross section

$$\text{Froude number Fr} = V/\sqrt{(g \times D)}, \quad \text{where } V = \text{Velocity of flow}$$

The free board (f) value is considered as per IS:10430-1982. The thickness for canal lining is considered as per Table 2.

Table 2 Thickness of canal lining

Sl No	Discharge (m^3/s)	Thickness (cm)
1	>5	7
2	2–5	6
3	<2	5

The Manning's equation for uniform flow is written as $Q = A\,\frac{1}{n} \times R^{\frac{2}{3}} \times S^{\frac{1}{2}}$, where Q = design discharge, S = longitudinal slope of canal. The discharge required by each canal is calculated on the basis of command area covered by the canal and is provided in Table 1. The design of canal of an irrigation project is always done from the tail end of the canal network as the main canal needs to carry the discharge required by the branch canals also. Due to long distance, main canals and branch canals containing distributary are designed as various subsections. The design discharge of canal sections of main canals D_1 and D_2 is provided in Tables 4 and 6, and the design discharge of branch canals and distributaries is given in Tables 5, 7 and 8, respectively (Table 3).

The longitudinal slope of various main canal sections is provided in Table 3, and slope for branch canals and distributaries is considered as 1 in 2800.

Table 3 Longitudinal slope for various main canal sections of Sukla IP

Canal	Chainage	Slope	Canal	Chainage	Slope
D_1	0.00–14,295 m	1 in 4000	D_2	0.00–7943 m	1 in 4100
	14,295–22,920 m	1 in 3800		7943–12,954 m	1 in 4100
	22,920–23,032 m	1 in 3100		12,954–26,385 m	1 in 4000
	23,032–24,932 m	1 in 3000		26,385–27,096 m	1 in 3000
	24,932–28,811 m	1 in 2800		27,096–31700 m	1 in 3000

4 The Nonlinear Optimization Model

Objective Function

The objective function of the nonlinear programming (NLP) model consists of cost of excavation and cost of lining to minimize the total cost of construction of the canal of

Table 4 Optimum dimensions and optimum cost of construction for main canal D_1

Sl no	Name	Discharge (m^3/s)	Optimum width B (m)	Optimum depth Y (m)	Optimum cost of construction per running meter (rupees)
1	D_1 from chainage 0.00 to 4024 m	11.709	4.524	2.262	7585.19
2	D_1 from chainage 4024 to 8107 m	11.284	4.462	2.231	7463.80
3	D_1 from chainage 8107 to 9418 m	9.915	4.250	2.125	6793.95
4	D_1 from chainage 9418 to 10271 m	9.418	4.169	2.085	6642.49
5	D_1 from chainage10,271 to 10,911 m	9.044	4.106	2.053	6525.06
6	D_1 from chainage 10,911 to 14,295 m	8.551	4.021	2.010	6369.10
7	D_1 from chainage 14,295 to 19,720 m	6.993	3.693	1.847	5779.70
8	D_1 from chainage 19,720 to 20,878 m	5.593	3.396	1.698	5263.91
9	D_1 from chainage 20,878 to 22,920 m	5.037	3.266	1.633	5042.00
10	D_1 from chainage 22,920 to 23,032 m	4.540	3.023	1.512	4012.87
11	D_1 from chainage 23,032 to 24,932 m	4.349	2.957	1.478	3914.50
12	D_1 from chainage 24,932 to 28,811 m	4.077	2.849	1.424	3756.96

Table 5 Optimum dimensions and optimum cost of construction for branch canals of D_1

Sl no	Name	Discharge (m^3/s)	Optimum width *B* (m)	Optimum depth *Y* (m)	Optimum cost of construction per running meter (rupees)
1	M_1D_1	0.425	1.220	0.610	1236.27
2	M_2D_1	1.37	1.893	0.946	2135.69
3	M_3D_1	0.497	1.294	0.647	1307.95
4	M_4D_1	0.374	1.163	0.582	1181.48
5	M_5D_1	0.493	1.290	0.645	1304.12
6	M_6D_1	1.457	1.937	0.968	2185.93
7	M_7D_1	0.102	0.715	0.357	772.91
8	M_8D_1	1.399	1.904	0.954	2152.61
9	M_9D_1	0.556	1.350	0.675	1362.68
10	$M_{10}D_1$	0.497	1.294	0.647	1307.95
11	$M_{11}D_1$	0.191	0.9040	0.452	940.79
12	$M_{12}D_1$	0.272	1.032	0.516	1058.24

Sukla Irrigation Project [6].

$$\begin{aligned}\text{Min } Z &= \text{Cost of excavation} + \text{Cost of lining} \\ &= C1 \times \text{Cross - sectional area of the canal} + C2 \times \text{Cross - sectional area of lining} \\ &= C1 \times B \times (Y + f) + C2 \times (B + 2 \times (Y + f)) \times t\end{aligned}$$

where

Z = Minimum cost of construction of canal per running meter.
$C1$ = Cost of excavation per cubic meter.
$C2$ = Cost of lining per cubic meter.

The values of $C1$ (item no. 6.1(A)) and $C2$ (item no. 18.4) are taken as per schedule of rates for the year 2019–20 of the irrigation department of Government of Assam.

Constraints

The various constraints of the optimization model are listed below:

i. *Discharge in the canal*: The discharge calculated from Manning's formula for the canal is greater than or equal to the design discharge for the canal.
ii. *Froude number*: The Froude number calculated for the canal section should be less than 1(one) to maintain the subcritical flow condition and to avoid hydraulic jump in the canal.

Table 6 Optimum dimensions and optimum cost of construction for main canal D_2

Sl no	Name	Discharge (m^3/s)	Optimum width (B) (m)	Optimum depth (Y) (m)	Optimum cost of construction per running meter (rupees)
1	D_2 from chainage 0.00 to 7943 m	12.711	4.687	2.344	7906.59
2	D_2 from chainage 7943 to 12,954 m	11.161	4.464	2.232	7468.30
3	D_2 from chainage 12,954 to 19,933 m	7.411	3.811	1.905	5989.05
4	D_2 from chainage 19,933 to 26,385 m	5.521	3.413	1.706	5291.63
5	D_2 from chainage 26,385 to 27,096 m	4.958	3.106	1.553	4135.75
6	D_2 from chainage 27,096 to 31,700 m	4.502	2.995	1.498	3971.41

Table 7 Optimum dimensions and optimum cost of construction for branch canals of D_2

Sl no	Name	Discharge (m^3/s)	Optimum width (B) (m)	Optimum depth (Y) (m)	Optimum cost of construction per running meter (rupees)
1	M_1D_2	1.55	1.982	0.991	2238.00
2	M_2D_2 from chainage 0.00 to 6492 m	3.75	2.761	1.380	3630.16

(*continued*)

Table 7 (*continued*)

Sl no	Name	Discharge (m^3/s)	Optimum width (B) (m)	Optimum depth (Y) (m)	Optimum cost of construction per running meter (rupees)
3	M_2D_2 from chainage 6492 to 13,716 m	2.718	2.447	1.223	3189.59
4	M_3D_2 from chainage 0.00 to 201 m	1.89	2.135	1.068	2416.17
5	M_3D_2 from chainage 201 to 1189 m	1.656	2.032	1.016	2295.47
6	M_3D_2 from chainage 1189 to 2042 m	1.244	1.825	0.913	2059.97
7	M_3D_2 from chainage 2042 to 3658 m	0.934	1.639	0.820	1657.52
8	M4D2	0.563	1.356	0.678	1368.96
9	M5D2	0.457	1.254	0.627	1268.87

Table 8 Optimum dimensions and optimum cost of construction for distributaries

Sl no	Name	Discharge (m^3/s)	Optimum width B (m)	Optimum depth Y (m)	Optimum cost of construction per running meter (rupees)
1	$S_1M_2D_2$	1.032	1.702	0.851	1923.18
2	$S_1M_3D_2$	0.234	0.976	0.488	1005.95
3	$S_2M_3D_2$	0.412	1.206	0.603	1222.65
4	$S_3M_3D_2$	0.31	1.084	0.542	1106.68

iii. *Condition of most economic rectangular canal section*: The condition for most economic rectangular canal section is $Y = \frac{B}{2}$. So, $B = 2Y$ is considered as a constraint to get the most economical canal section.
iv. *Width of the channel*: The width of the channel cannot be negative, i.e., $B > 0$. Considering the construction point of view, let the minimum width of the channel be 0.1 m.
v. *Depth of the channel*: The depth of the channel cannot be negative, i.e., $Y > 0$. Let the minimum depth of the channel be 0.1 m.
vi. *Limiting velocity*: The limiting velocity for the canal is considered as per IS 10430:2000, and for cement concrete lining, it is 2.7 m/s [5].
vii. *Critical velocity*: According to IS 10430:2000, the critical velocity obtained by any formula should be less than the velocity of the canal obtained by Manning's formula. The critical velocity obtained by Kennedy's formula, i.e., $V_0 = 0.546 \times D^{0.64}$, should be less than the velocity in the canal obtained by Manning's formula [5].
viii. *Top width*: The max top width for main canal and branch canal is 5 m and 3 m, respectively.

Variables

The optimization model is subjected to two variables. They are: (i) bed width of the canal and (ii) depth of the canal.

5 Results and Discussions

The nonlinear optimization model is implemented for the canal network of Sukla IP, and optimum dimensions and optimum construction cost for the canal sections per running meter are obtained. The optimum dimensions and optimum construction cost for main canal D_1 and branch canals of D_1 are given in Tables 4 and 5, respectively. For sections of main canal D_2, branch canals of D_2 and distributaries of D_2, optimum dimensions and optimum construction cost are provided in Tables 6, 7 and 8, respectively. The total minimum construction cost for the whole canal system will be Rs. 50,65,06,314.00 (rupees fifty crore sixty-five lakh six thousand three hundred and fourteen only).

6 Conclusions

This paper presents a nonlinear optimization model for optimal design of rectangular canal section of Sukla Irrigation Project which is located in Baksa district of Assam, India. The optimization model is solved in Microsoft Excel solver GRG nonlinear engine, and minimum cost of construction for the whole canal system of Sukla IP is obtained. The most economical canal dimensions are also determined.

References

1. Bhattacharjya RK (2006) Optimal design of open channel section incorporating critical flow condition. J Irrig Drain Eng 132(5):513–518
2. Bhattacharjya RK, Satish MG (2007) Optimal design of a stable trapezoidal channel section using hybrid optimization techniques. J Irrig Drain Eng 133(4):323–329
3. Chow VT (1959) Open channel hydraulics. E-book, McGraw-Hill, New York
4. Hameed AT (2010) Optimal design of round bottomed triangle channels. Tikrit J Eng Sci 17(3):31–43
5. IS 10430:2000 Indian Standard. Criteria for design of lined canals and guidelines for selection of type of lining. Bureau of Indian Standards, New Delhi
6. Schedule of Rates (civil parts) for the year 2019–2020. Irrigation Department, Government of Assam
7. Swamee PK et al (2000) Comprehensive design of minimum cost irrigation canal sections. J Irrig Drain Eng 126(5):322–327

River Management

Chapter 14
Comparison of Simple and Modified SCS-CN in Runoff Prediction in a Highly Flood Prone Zone

Nameirakpam Momo Singh, Thronlem Winkangshu, and Thiyam Tamphasana Devi(✉)

Department of Civil Engineering, National Institute of Technology, Manipur, Imphal, Manipur, India

Abstract. Imphal West District (area = 558.00 km^2) in Manipur, which is within the Imphal River Basin, has witnessed severe flood problems recently due to continuous rainfall events in the year 2015, 2016, and 2017. With the help of satellite imagery and geospatial techniques, modeling of runoff in this region has been carried so that trend of present and future intensity of flood can be understood. Soil conservation services curved number (SCN-CN) method is a widely accepted and used method in modeling runoff. But one important limitation of this method is, in this method, slope is not considered and hence not taken into account while developing the modeling parameters. Therefore, in this study, runoff volume is modeled and predicted by the simple and modified SCS-CN method and results are compared. The model parameters considered in SCS-CN were hydrological soil group (HSG) which is the most concerning factor, types of landuse and landcover (LU/LC), and the other influencing factor (rainfall for the year 2007–2017) of HSG required to derive the curve number (CN). But in the modified SCS-CN method, additional affecting parameter slope is considered to enhance the accuracy of modeled results. The GIS tools ArcGIS® and ERDAS® were used to generate the slope and landuse/landcover (LU/LC) of the study area and as the end product of remote sensing, the satellite image, Landsat 8 OLI is used in this study. In both the results, the highest amount of runoff volume is evident in the year 2010 and 2016 and the lowest in 2008. It is concluded that the SCS-CN method integrated with remote sensing (RS) and geographical information system (GIS) may serve as a reliable tool for hydrological modeling with the limited available hydrological parameters.

Keywords: SCS-CN · Modified SCS-CN · Imphal West · LU/LC · RS and GIS · HSG

1 Introduction

Runoff is one of the most important hydrologic variables used in most water resources applications. It is considered to be the major factor for the flood which is due to the

R. K. Bhattacharjya et al. (eds.), *Sustainable Water Resources Management*,
Advances in Sustainability Science and Technology,
https://doi.org/10.1007/978-981-16-7535-5_14

sudden high-intensity rainfall or prolong non-stop rainfall events. Apart from these rainfall characteristics, there is a number of catchment-specific factors, which have a direct effect on the occurrence and volume of runoff. This includes soil type, vegetation cover, and slope and catchment type. Runoff modeling is done by using soil conservation service curve number (SCS-CN) method which is a widely accepted model by several researchers and hydrologists. The runoff model is best expressed by spatially variable parameters such as rainfall, soil types and land use/land cover, etc. [18]. To successfully adopt the SCS-CN method [11], computer-based techniques, i.e., geospatial information system (GIS) and satellite images (which contain areal and topographic information of earth and its surroundings) are applied together [12]. For efficient management of large and complex databases, a GIS which is a computer-based tool that displays, stores, analyzes, retrieves, and generates spatial and non-spatial (attribute) data provides suitable alternatives [1]. However, a major limitation of this method is that it does not take into account the effect of slope, and therefore, modified SCS-CN [2, 9, 15, 20] is introduced which considers the effect of slope in runoff modeling.

In this study, with the help of satellite imagery and geospatial techniques, runoff modeling in a highly flood-prone zone (Imphal West District in Manipur) is carried out by using the simple and modified SCS-CN method. Issues and concerns in the selected study area are that the study region is in the valley region of the state which is surrounded by 98% of mountains [10]. Rivers from these hills flow into the valley and very often lead to flash floods every year. Thus, river flooding is a regular hazard faced by the state in general during the monsoon season. All the major river (Imphal, Iris, and Thoubal) systems in the state are vulnerable to flooding, as captured in the Vulnerability Atlas. The urban areas face flooding primarily due to drainage failures and increased runoff loads on hard surfaces. The valley district in Manipur namely Imphal East, Imphal West, Thoubal, and Bishnupur are most vulnerable to flood which is primarily due to heavy rainfall. In these areas erosion, sliding, and slumping of the banks are common during the rainy season.

2 Study Area

Imphal West district is one of the sixteen districts of Manipur state in northeastern India which is shown in Fig. 1a and the drainage pattern of the study area is shown in Fig. 1b. The district falls in the category of the Manipur valley region. It is a tiny plain at the center of Manipur surrounded by plains of other districts. As of 2011, it is the most populous district in the state, with an average elevation of 786 m (2579 ft). Imphal West District shares its boundaries with Senapati District in North, Thoubal District and Bishnupur District in the south, Imphal East District and Thoubal district in East, and Senapati District and Bishnupur District in West. It lies between Coordinates—24.9°N to 93.9°E having a total area of 558 km^2. The average rainfall in this district is around 1581 mm and temperature varies between 11 and 34°C.

3 Materials and Methods

The data required for this study are rainfall data, digital elevation model (DEM) data, LANDSAT data, and soil data. The source and description of data used in this study are (i)

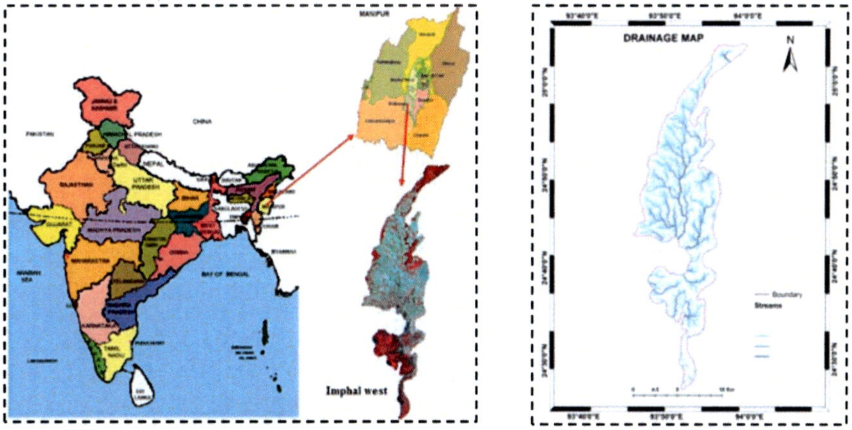

(a) Imphal West **(b) Drainage pattern**

Fig. 1 **a** Location of study area (Imphal West) and **b** drainage pattern

for spatial, i.e., SRTM DEM, Landsat 8 Image (30 × 30 m resolution) (USGS) OLI/TIRS C1 Level-1 and (ii) for non-spatial data, i.e., rainfall (Indian Council of Agricultural Research, ICAR, Imphal), soil (National Bureau of Soil Survey and Land Use Planning, NBSS & LUP, Nagpur). The GIS tools used in this study are ArcGIS 10.2® and ERDAS IMAGINE®. ArcGIS 10.2® is a geographic information system (GIS) for working in an integrated environment with maps and geographic information to create and analyze various sets of the database. ERDAS IMAGINE® is an image processing software package that allows users to process geospatial and various imagery in raster and vector format. In the modeling of runoff volume in this study, the basic theoretical formulae of SCS-CN and its modified version were used. The role of RS and GIS were involved in the determination of CN values which again depends on soil type, hydrological soil group, and landuse/land cover (LU/LC) type. And details of this approach are presented in Sect. 3.1.

Calculation of Runoff Volume

Simple SCS-CN Method

Runoff volume by CSC-CN method is expressed as:

$$Q = \frac{(P - 0.3S)^2}{(P - 0.7S)} \tag{1}$$

where Q = Surface runoff depth (mm) and volume of runoff volume in m^3 is calculated multiplying by basin area, P = precipitation (mm), S = potential maximum soil moisture retention (mm). The potential maximum retention S (mm) can vary in the range of $0 \leq S \leq \infty$, and it is directly linked to curve number (CN). So, S is given as:

$$S = \frac{25400}{\mathrm{CN}} - 254 \tag{2}$$

where CN = Curve Number, is a function of LU/LC, soil type, HSG (Type A, B, C, and D), and antecedent moisture condition (AMC), and it can vary from 0 to 100. Higher AMC and CN values would indicate more runoff potential and vice versa; therefore, median CN obtained from an array of CN values would commonly be adopted for the watershed [6, 7, 16, 17]. For Indian conditions, some modifications were done by the soil and water conservation Department, Ministry of Agriculture, New Delhi. For Indian conditions, $I_a = 0.3S$ [13, 14]. For different types of CN values, weighted CN is calculated by applying the basic principle of weighted value calculation.

Modified SCS-CN Method

The simple SCS-CN method does not take into account the effect of slope on runoff yield. However, there are few models which incorporate a slope factor to the CN method to improve the estimation of surface runoff depth and volume [8]. Fewer attempts have been made to include the slope factor into the CN method. Those who had taken the slope factor into account were notably, [8, 19]. The following equation was used to adjust the CN values obtained from SCS-CN standard tables for the slope, respectively. Both methods assume very simply that CN obtained from SCS standard tables correspond to a slope of 5%.

$$\mathrm{CN}_2\alpha = \mathrm{CN}_2 \times \mathrm{K} \tag{3}$$

where $\mathrm{CN}_2\alpha$ is the value of CN_2 for a given slope α, CN constant $K = (322.79 + 15.56\alpha)/(\alpha + 323.52)$.

4 Results and Discussion

Generated Soil Map

The delineated soil map for Imphal West district from the soil map of the state which is collected from NBSS, Nagpur is shown in Fig. 2. Then boundaries of the different soil groups were digitized and the vector map of the soil was prepared. The spatial distribution of different classes of soil for Imphal West is characterized by fine loamy clayey in texture having a serious problem of drainage in addition to frequent flooding. They are clayey and silty both at the surface and in the subsoils.

Land Use/Land Cover Classification

Supervised classification of LU/LC was done and is presented in Fig. 3. Six different categories, i.e., water bodies, forest, agriculture, built-up, and wetlands were classified. Accuracy assessment of this classified LU/LC was performed, and the overall accuracy was found to be above 90% in the higher range with a Kappa coefficient of 0.95 which is greater than 0.75 and is excellent.

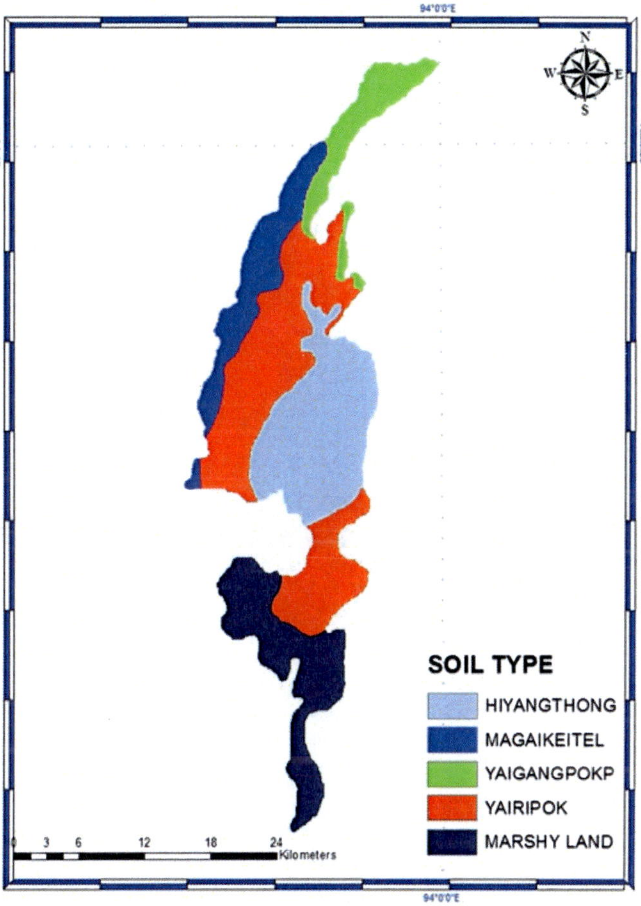

Fig. 2 Soil map of Imphal West

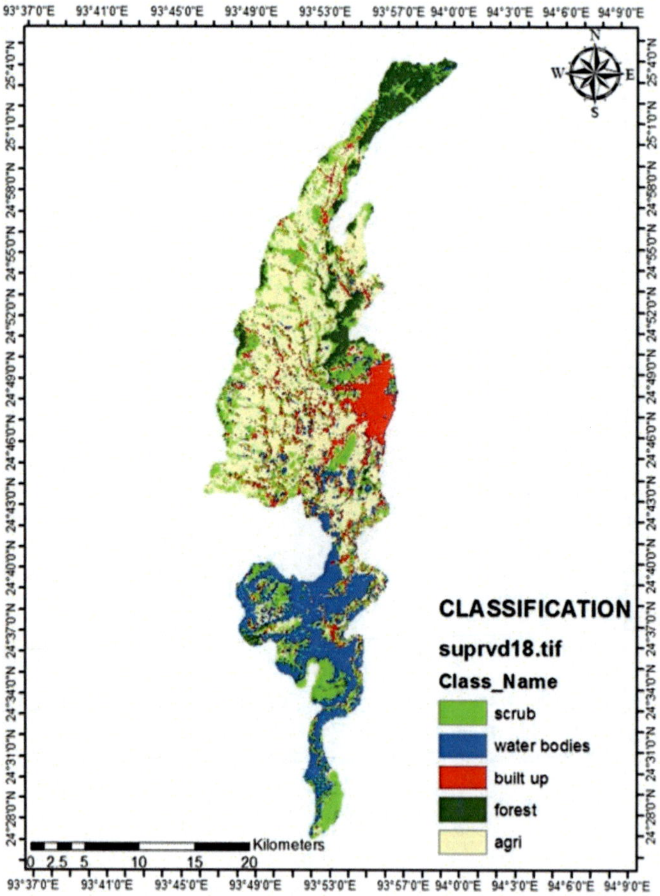

Fig. 3 LULC of Imphal West by supervised classification

Simple CN

The CN value for each soil hydrologic group and corresponding landuse class are presented in Table 1. It is observed from this study that hydrologic soil group B leading to low CN value while the hydrologic group D leading to a high CN value in the catchment. Gandini and Usunoff [5] observed that hydrologic soil group B leading to lower CN values in a humid temperate area. The CN values map is shown in Fig. 4. Based on the soil properties, the overlay of the LU/LC map and soil map generates the CN for different LU/LC. Using the CN generated, S is obtained after which the initial abstraction (I_a) is obtained. Finally, the runoff is estimated for different LU/LC which is shown in Table 2 where the comparison is made with the values predicted by the modified SCS-CN method.

Table 1 Curve Number for different land use/land cover

Sl. no	LULC	CN
1	Forest	60
2	Agriculture	89
3	Scrubs	73
4	Built up	78
5	Water bodies	97

Modified CN and the Final Result

To incorporate the factor of geomorphology into account for the computation of modified CN, the slope map (Fig. 5) provides the slope of the area in terms of degree, which then is converted into intercepts to compute the slope factor 'K.' Then based on the above-mentioned methods and approaches, annual average runoff volume for the year 2007–2017 is estimated by using the simple and modified SCS-CN method which is tabulated in Table 2.

In both cases, the highest runoff volume is found to be in the years 2010 and 2016 followed by in the years 2013 and 2015, and the lowest observed in the year 2008. So, this result somewhat supports the actual flood that happened in the year 2015 [21] 2016 [4] & 2017 [3]. There is no significant deviation in the estimated values of these two approaches, however, the modified SCS-CN method under-predicts slightly (at the maximum of about 1.5%) from the Simple SCS-CN method. A column chart of the comparison of estimated runoff volume by these methods is shown in Fig. 6.

5 Conclusions

This study reflects and supports the actual flood happened in the year 2015–2017 which predicts highest runoff volume by both the methods. Furthermore, this study gives the

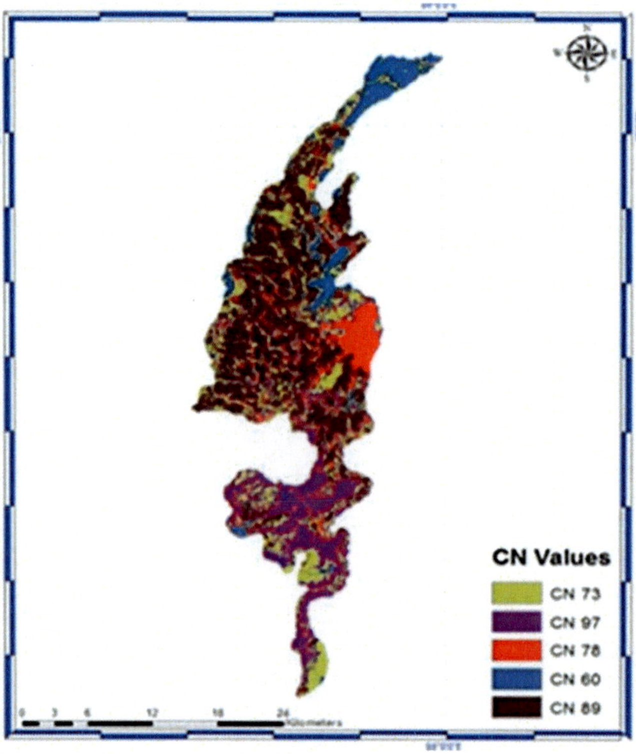

Fig. 4 CN values map

Table 2 Runoff for SmCS-CN and Modified methods

Year	SCS-CN method (m^3)	Modified SCS-CN method (m^3)
2007	349,279.7	346,717.7
2008	249,794.0	248,043.1
2009	405,837.6	401,102.3
2010	554,309.1	546,863.4
2011	468,200.2	461,329.8
2012	364,378.9	360,080.6
2013	525,819.4	519,039
2014	439,614.5	433,488.3
2015	523,032.0	515,099.5
2016	561,736.5	554,560.8
2017	492,615.7	485,194

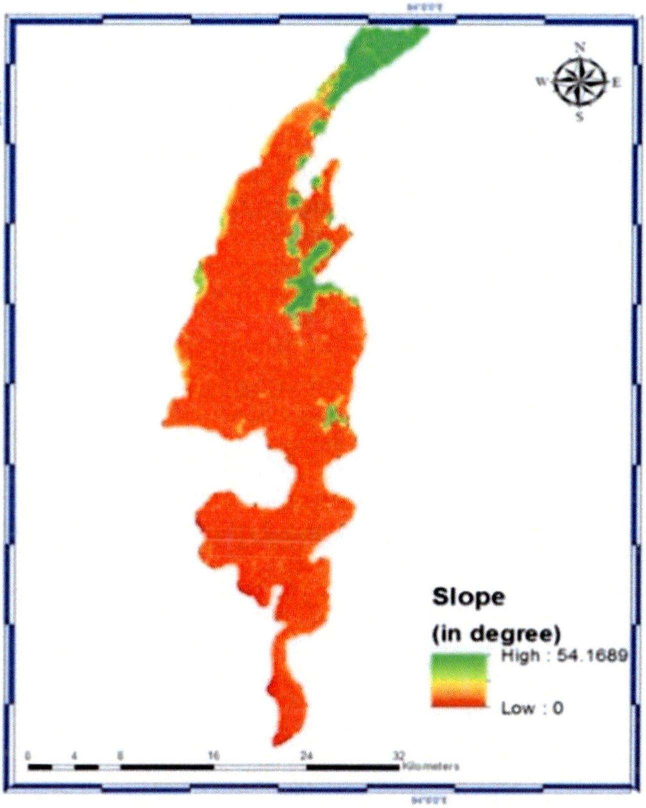

Fig. 5 Slope map

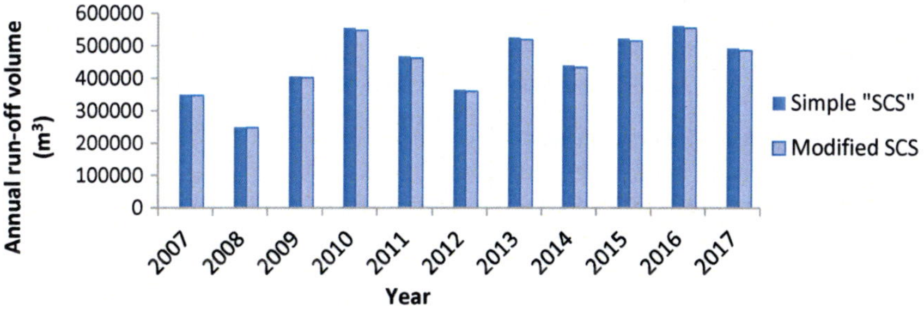

Fig. 6 Comparison of runoff volume predicted by simple and modified SCS-CN method

water yield estimation which can be useful for the development project in the watershed. Thus, it has been demonstrated that GIS and remote sensing techniques can provide reliable and significant information required by SCS-CN, and analytical capability to water resource assessment of a certain basin even though there are no enough available data. Hence, further studies on this basin could make use of this versatile tool to update the hydrologic description of the area. It concludes with RS and GIS-based SCS-CN method as an efficient method for computing runoff and can be further used for planning and management of the available water resources.

References

1. Arwa DO (2001) GIS-based rainfall-runoff model for the Turasha Sub Catchment Kenya. MSc thesis, International Institute for Aerospace Survey and Earth Sciences, Enschede, The Netherlands
2. Costa FFD, Paz ARD (2019) Modified NRCS-CN-TUH methods for distributed rainfall-runoff modelling. RBRH (24), Porto Alegre, Epub
3. Directorate of Fisheries Manipur (2018) Annual action plan on "disaster management plan 2017–18", p 9
4. Epao.net (local daily e-newspaper published on 22 May 2016). http://www.e-pao.net/epGallery.asp?id=2&src=News_Related/Calamities_News_Gallery/Flood20160522_1
5. Gandin ML, Usunoff EJ (2004) SCS curve number estimation using remote sensing NDVI in a GIS environment. J Environ Hydrol 12:16
6. Hawkins RH, Hjelmfelt AT Jr, Zevenbergen AW (1985) Runoff probability, storm depth and curve numbers. J Irrig Drain Eng ASCE 111(4):330–340
7. Hjelmfelt AT Jr (1991) Investigation of curve number procedure. J Hydraul Eng 117(6):725–737
8. Huang GH, Zhang RD, Haung QZ (2006) Modelling soil water retention curve with fractal method. Pedosphere 16(2):137–146
9. Kabiri R (2014) Simulation of runoff using modified SCS-CN method using GIS system, case study: Klang watershed in Malaysia. Res J Environ Sci 8:178–192
10. Laiphrakpam L, Das AK, Singh PK, Ghosh SK (2014) Studies on orchid flora of Imphal valley, Manipur. Indian J Appl Pure Biol 29:9–23
11. McCuen RH (2002). Approach to confidence interval estimation for curve numbers. J Hydrol Eng 7(1):43–48. https://doi.org/10.1061/(ASCE)1084-0699(2002)7:1(43)
12. Mohan, Shrestha MN (2000) A GIS based integrated model for assessment of hydrological change due to land use modifications, proceeding of symposium on restoration of lakes and wetlands. Indian Institute of Science, 27–29 November, Bangalore, India
13. Narayana DVV (2002) Soil and water conservation research in India. Indian Council of Agricultural Research, New Delhi
14. Rao KN, Narendra K, Latha PS (2010) An integrated study of geospatial information technologies for surface runoff estimation in an agricultural watershed, India. J Indian Soc Remote Sens 3:255–267
15. Satheeshkumar S, Venkateswaran S, Kannan R (2017) Rainfall–runoff estimation using SCS–CN and GIS approach in the Pappiredipatti watershed of the Vaniyar sub basin, South India. Model Earth Syst Environ 3(24):1–8
16. Schneider LE, McCuen RH (2005) Statistical guidelines for curve number generation. J Irrig Drain Eng 131(3):282–290

17. SCS (1954, 1956, 1964, 1965, 1971, 1972, 1985, 1986, 1993, 1997) Hydrology, National Engineering and book, supplement A, section 4, chapter 10. Soil Conservation Service, USDA, Washington, DC
18. Sharma D, Kumar V (2002) Application of SCS model with GIS data base for estimation of runoff in an arid watershed. J Soil Water Conserv 30(2):141–145
19. Sharpley AN, Williams JR (1990) EPIC-erosion/productivity impact calculator: 1. model documentation. Technical Bulletin No.1768, US Department of Agriculture, US Government Printing Office, Washington, DC
20. Viji R, Prasanna PR, Ilangovan R (2015) Modified SCS-CN and green-Ampt methods in surface runoff modelling for the Kundahpallam Watershed, Nilgiris, Western Ghats, India. Aquat Proc 4:677–684
21. Zutshi B (2018) Manipur flood 2015: preparedness and disaster risk reduction strategies. In: Development and disaster management, pp 65–81

Chapter 15
Quantification of Discharge Hysteresis Produced in Amazon River Basin

Durga Sharma[1(✉)] and Basudev Biswal[2,3]

[1] Department of Civil Engineering, IIT Hyderabad, Sangareddy, India
[2] Department of Civil Engineering, IIT Bombay, Mumbai, India
[3] The Interdisciplinary Programme (IDP) in Climate Studies, IIT Bombay, Mumbai, India

Abstract. Hysteresis in hydrology has been observed in many areas like for instance, storage discharge relationship, soil–moisture relationship and sediment discharge relationship. It is extremely important to quantify these hysterical patterns properly and effectively. Many indices have already been used by researcher to quantify hysterical pattern, but most of them lack proper range to quantify the pattern. Here in this paper, we have developed a hysterical index which ranges from 0 to 1. The hysterical index so determined is based on analysing two sine curves, if one is denoted by $\sin(x)$, the other is considered as $\sin(x + t)$. This analysis has been done for basin having area ranging from few 100 km^2 to 6.65 × km^2. The basin having more area shows hysterical pattern with hysterical index varying from 0.1 to 0.89. The highest hysterical index was found for Obidos River basin in Amazon River. Catchments with largest area in Mississippi River basin have hysterical pattern for less lag time. The hysterical index for Mississippi River basin was found around 0.23. In some of the basins, no hysteresis were formed, so for them hysterical index is not determined as it will be close to zero.

Keywords: Amazon river discharge · Hysterical quantification · Autocorrelation · Streamflow

1 Introduction

Hysteresis is where output depends not only on the current state of a system but also on the input history and storage trajectory in time [1, 2]. In hydrology, hysteresis has been observed in storage discharge relation [1, 3], soil–moisture hysteresis and sediment discharge hysteresis [4–6]. It is now recognized as the collective name for strongly nonlinear phenomena [7]. Rate-dependent hysteresis has been observed in hydrology for a long time in the looped rating curve and in similar plots of state and rate variables from systems of linear and nonlinear differential equations that model hydrological processes [8]. The hysteresis formed between these parameters is often quantified by hysteresis index, for instance, for sediment discharge hysteresis, Langlois et al. developed a hysteresis index by splitting the event hydrograph in a rising and falling limb and by

R. K. Bhattacharjya et al. (eds.), *Sustainable Water Resources Management*, Advances in Sustainability Science and Technology,
https://doi.org/10.1007/978-981-16-7535-5_15

computing the regression lines for both sub-data sets. Further, another index developed by Lawler et al. [9] is based on the ratio of n suspended sediment concentration values at the mid-point, between the minimum discharge (Qmin) and the maximum discharge (Qmax) of an event. Here in this paper, we will see hysteresis formation between streamflow (Q_t) and past ($Q_{t\text{-}\mu}$), where Q_t *is* a discharge at time *t* and $Q_{t\text{-}\mu}$ is the past discharge which is μ days before the time *t*. This analysis has been carried at different catchment scales with areas varying from a few 100 km^2 to 6.65×10^6 km^2. This study has been carried out in the sub-basins of Amazon and USGS basins. The hydrology of the Amazon River Basin, which has about 20% of global river runoff and a massive flood plain system, has always been a great area of study and a fascinating topic among researchers and hydrologists. It is the world's largest river basin, with an area of about 6.65×10^6 km^2 and average annual rainfall of 2200 mm a year (Tropical Rainfall Monitoring Mission (TRMM)). It gathers its waters from five degrees north latitude to 20 degrees south latitude. It also holds the world's largest stand of tropical rainforest, provides a home to about one-quarter of all terrestrial species on earth [10] and accounts for ~15% of global terrestrial photosynthesis [10]. While the forest dynamics and the fate of the Amazon remain poorly understood [10, 11], climate models indicate that moisture stress will be a dominant feature, particularly in the southern and eastern parts [10, 11]. However, the climate models used for future projections treat soil water rather simplistically; soil column is 2–3 m deep and freely drained. So, research on actual requirements for a numerically consistent representation of flow dynamics in large-scale river-flood models is needed to improve both modelling performance and computational efficiency [12]. And further, it is extremely important to find the degree of hydrologic alteration in a stream so that water resource managers can develop better understanding of human effects on streamflow and establishment of ecological flow recommendations for the sustainability of a stream's biological health can be determined [13–16]. Not only for these issues, degree of hydrological variation is extremely important for nutrition loss prediction for agricultural catchment and flood drought frequency analysis. We have hydrological indices quantifying degree of hydrological changes in river system such as flashiness index [17]. Here in this study, we explore whether only past discharge can be used to predict the discharge at different lead time or not, and we will come up with some interesting hysterical pattern observed in these river basins and will develop hysterical index to quantify the hysterical pattern.

2 Material and Method

Flow Data Acquisition

We have used streamflow data of Amazon River basins and US basins. Discharge data for Amazon River basin and its tributaries have been downloaded from ORE (Environmental Research Observatory) HYBAM (Geodynamical, hydrological and biogeochemical control of erosion/alteration and material transport in the Amazon River basin—http://www.ore-hybam.org/). Mean daily streamflow data were obtained from the US Geological Survey—https://waterwatch.usgs.gov/ for 25 stream gauge sites from Mississippi, Missouri, Kansas and Colorado basins. Streams were selected based on the availability of daily records of at least 30 years. Then streamflow data analysis was done as method discussed below.

Autocorrelation Analysis of Discharge Data

The autocorrelation function is the expected value of a product. Assume we have a pair of random variables from the same process, $X1 = X\ (t1)$ and $X2 = X\ (t2)$, then the autocorrelation is often written as

$$R(t1, t2) = \frac{E[(X_{t1} - \mu_{t1})(X_{t2} - \mu_{t2})}{\sigma_{t1}\sigma_{t2}} \tag{1}$$

The value of R as represented in Eq. (1) must lie in the range $[-1, 1]$, with 1 indicating perfect correlation and -1 indicating perfect anti-correlation. For Amazon River basin at Obidos, it varies from -0.8 to 1 as shown in Fig. 4b . Then in order to see how discharge data are autocorrelated, we used expression (2) for analysis.

$$Q_t = k_\mu Q_{t-\mu} \tag{2}$$

Here Q_t represents discharge at any time t, $Q_{t-\mu}$ is the discharge lag discharge, and μ is the lag time period in days. This expression can be used for discharge prediction for large river basins. From Fig. 1, *Q*pred is Q_t and *Q*obs is the past observed discharge, i.e. μ days before.

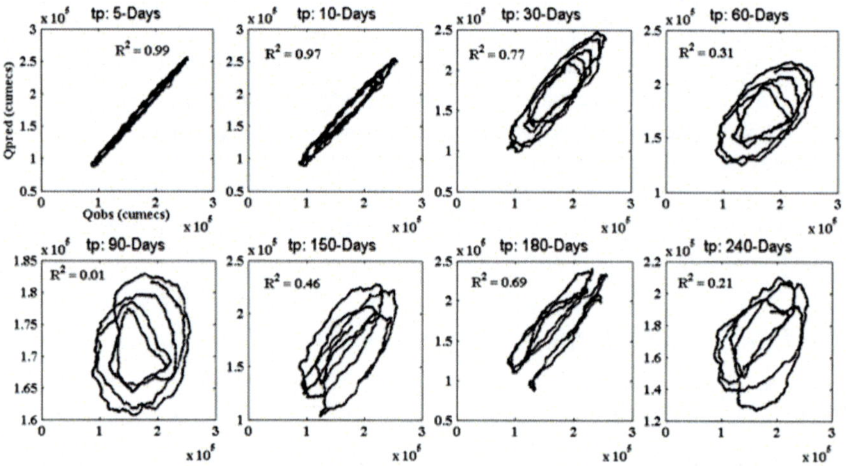

Fig. 1 Hysterical formation at different lag time with the scatter plot between *Q*Pred and *Q*obs at Obidos gauging station. *tp* represents past time or lag time in days

Hysterical Quantification of Discharge Data

The interpretation of hysterical curves is required to understand on what factor the hysterical forming quantities depend on, how they vary depending on the characteristics and size of a catchment. Quantitative descriptions of the hysteretic loop have been discussed in the literature, and various methods of computing hysteresis indices (HI) have been proposed as seen in Butturini et al. [18], for example using the relative difference between extreme concentration values or using the ratio of turbidity values in rising and falling limbs of the storm hydrograph at the mid-point discharge value [9]. Further in order to obtain a hysteresis index that is not influenced by the absolute quantity of discharge (Q) and suspended sediment concentration (SSC), [4] have normalized both measures. In most of the methods for obtaining hysterical quantification, the limit is not defined. Here in our paper, hysterical quantification between discharge and past discharge will vary from 0 to 1. If the parameters are showing good hysterical pattern, then we will have value of its hysterical index as 1, if however, the hysterical patterns are not good at all, then the hysterical index assigned to it will be 0. We have quantified hysterical pattern based on sine curve analysis. We have noticed that generally, the hysterical curves follow the sine curve or cosine curve. We tried to make a simple hysterical equation in the form of sine curve, the discharge of Amazon River basin due to its seasonality shows a good sine curve pattern. Further, two sine curves, when plotted with respect to each other, form hysteresis as shown in Fig. 3. Then we determined the correlation between two halves of a complete sine curve. The same varied from 0 to 1. Similar analysis was done for basins in Amazon, Mississippi, Missouri and Colorado (Table 1).

3 Result and Discussion

Discharge prediction using Eq. (2) was also done for some river basins as mentioned in Table 1. Prediction was done for lag values of 5, 10, 30, 60, 90, 150, 180 and 240 days. Linear regression analysis was done to find constant k in Eq. 1. Then scatter plot was between predicted discharge data and observed discharge data plotted, and we observed hysterical pattern in most of the Amazon River basin. One example has been shown in Fig. 1. Coefficient of correlation decreases continuously with increase in lag time till 90 days and then increases from 150 days, and at 180 days, coefficient of correlation is 0.69. For Amazon River basin, coefficient of correlation was found to be 0.99, 0.97, 0.77, 0.31, 0.01, 0.46, 0.69 and 0.21 for lag days of 5, 10, 30, 60, 90, 150, 180 and 240, respectively. Similar analysis was done for basins in Mississippi, Missouri, in USA. We selected 25 gauging station in USA. The largest area of US basin of Mississippi showed hysterical pattern for lag days 5, 10 days (see Fig. 2). Further since hysteresis was obtained while plotting scatter plot (Fig. 1), we quantified hysterical pattern.

Hysterical pattern was observed in most of the Amazonian basins, and the hysterical index varied from 0.1 to 0.89 as shown in Table 1. Very few hysterical pattern were noticed in USA basins even if the area was large. Further yearly one hysteresis is occurring in most of Amazon River basins, and they follow pattern as shown in Fig. 4a. Formation of hysteresis is due to high autocorrelation of streamflow; the same has been shown in Fig. 4b. The figure shows autocorrelation factor for gauging station Obidos varying from 0.99 to −0.8. Well-defined hysteresis was formed in Amazon basins. Also it has

Table 1 Hysterical index of river basins

Station name (Amazon)	Stations Id	Hysterical index	USA basins Id	Hysterical index
Obidos	17,050,001	0.89	7,289,000	0.24
Manicore	15,700,000	0.78	7,010,000	0.21
Humaita	15,630,000	0.72	07,020,500	0.21
Porto	15,400,000	0.67	05,587,500	0.14
Abuna	15,320,002	0.65	06,892,350	0.11
Manacapuru	14,100,000	0.58	06,774,000	0.14
Itapeua	13,150,000	0.49	13,213,100	0.17
Santo	11,500,000	0.17	5,378,500	No hysteresis
Sao	11,400,000	0.12	5,331,000	No hysteresis
Tabatinga	10,100,000	0.14	8,150,000	No hysteresis
Fortaleza	17,500,000	0.79	07,022,000	No hysteresis
Itaibuba	17,730,000	0.71	8,150,700	No hysteresis
Jatoba	17,650,000	0.68	11,204,100	No hysteresis
Barra	17,430,000	0.63	11,507,500	No hysteresis
Luccas	17,230,000	0.58	7,020,500	No hysteresis
Canutam	13,880,000	0.78	9,402,500	No hysteresis
Labrea	13,870,000	0.74	6,177,000	No hysteresis
Valparaiso	13,710,001	0.52	07,335,500	No hysteresis
Seringal	13,750,000	0.42	08,162,000	No hysteresis
Monoel	13,180,000	0.35	03,611,000	No hysteresis

First column gives the names of the Amazon River basins used in the analysis. Second column is the gauging station id of the corresponding basins for Amazon River basins shown in column 1. Third column is the hysterical index varying from 0 to 1. Fourth column is the USGS river basins id. Fifth column gives information about hysterical index of the USGS River basins

been noticed that hysteresis formed is rate-dependent hysteresis as the shape of the curve changes with change in lag duration (Figs. 1 and 2). Deviation in hysterical behaviour is observed if there is difference in rainfall pattern in the consecutive years. Hysterical quantification was done if the river basin forms hysteresis. In the US basin, hysterical pattern was observed for few basins, which were larger in area. Also, the lag period was very less, i.e. 5 and 10 days (refer Fig. 2). The hysterical index for those basins was small as compared to basins in Amazon with value 0.23. In Amazon River basin, high hysterical index is due to less human interference over Amazon River basin, frequent rainfall and larger lag time. The reason behind less value of hysterical index and no hysterical formation in Mississippi river is due to less rainforest area, high anthropogenic effect, etc. Further if we plot hydrograph at Amazon River basin and at one of the gauging stations at Mississippi River basin, we get smooth curve for Amazon basin and highly

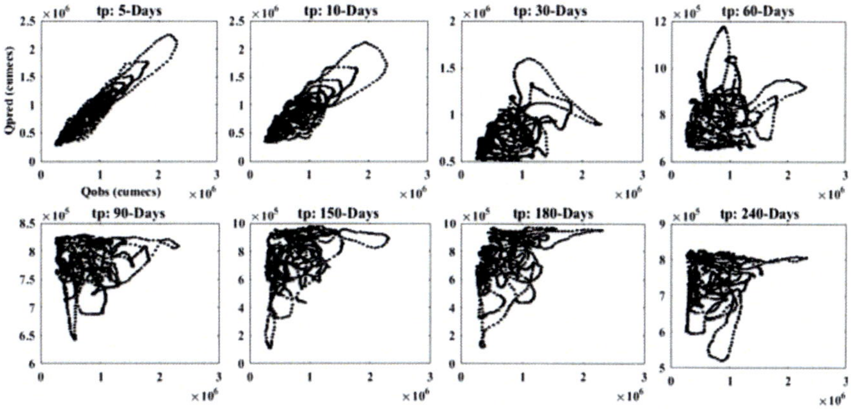

Fig. 2 Hysterical formation at different lag time shown in the scatter plots between *Q*Pred and *Q*obs at Mississipi (basin id: 7,289,000) gauging station. *tp* represents past time or lag time in days

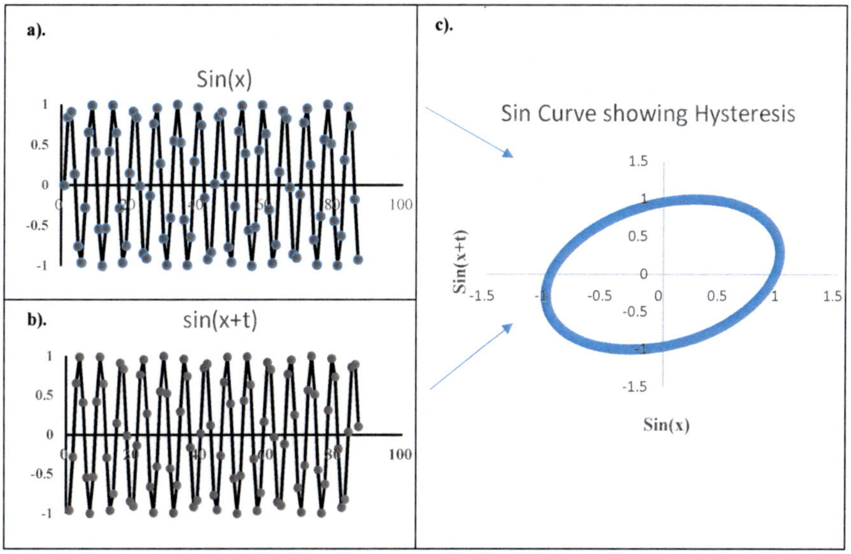

Fig. 3 Hysterical analysis using sine curve with lag time of t. **a** Plot shows sine curve for function $f(x)$. **b** Plot shows sine curve for function $f(x+t)$, where t is the lag time. **c** Scatter plot of $\sin(x)$ and $\sin(x+t)$ showing hysterical pattern

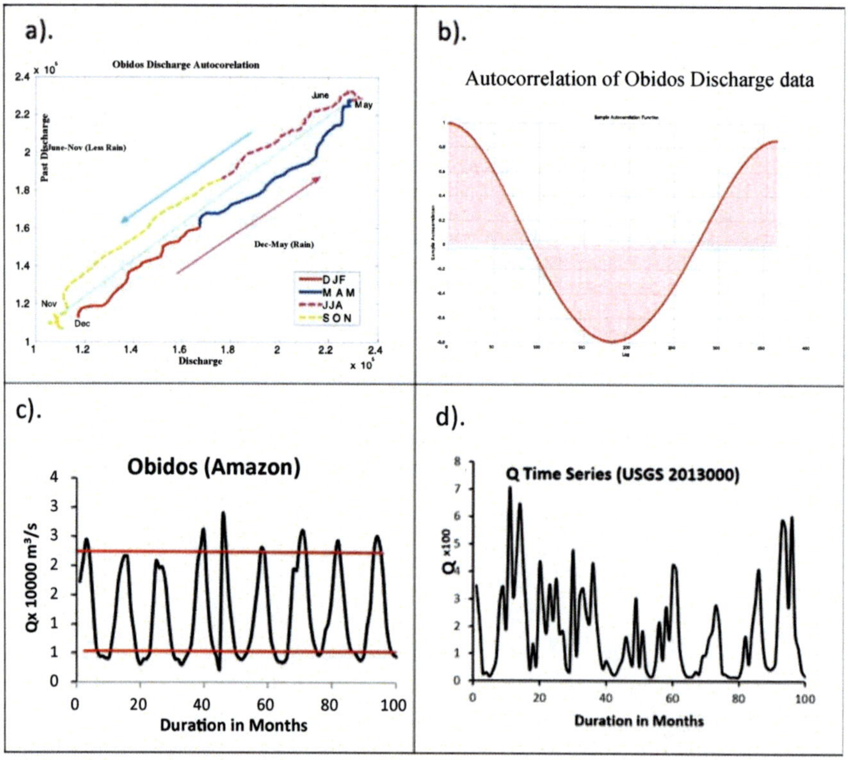

Fig. 4 Hysterical analysis: **a** Yearly hysterical pattern of streamflow for Amazon River at Obidos, discharge on both axes are in m^3/s. **b** Autocorrelation of streamflow at Obidos. **c** Hydrograph showing smooth curve for Obidos. **d** Hydrograph showing flashy behaviour of basin 20,130,000 (USGS basin)

flashy curve for Mississippi River basin as shown in Fig. 4c, d respectively. Then we analysed hysterical index with catchment area. Scatter plot between area and hysterical index showed no significant correlation; however as the area decreases, the hysterical index decreases.

4 Conclusion

Most of the discharge prediction equations are complicated and require lot of data and catchment information. If the streamflow of river basins is showing good autocorrelation, then discharge prediction in those catchments can be done using past discharge and hysterical index factor. For quantifying hysterical index, many methods are there; here in this paper we have defined a new technique of finding hysterical index which varies from 0 to 1. The highly hysterical streamflow will have hysterical index as 1, and streamflow having no hysterical curves will have 0 hysterical index. We have observed that these types of phenomena are generally observed in basins having larger area. Amazon basin has highest hysterical index of 0.89 at Obidos. And this value decreases as the basin area decreases. No particular relationship has been established between basin area and hysterical Index. Hysterical index for other river basins from Amazon and the USA has been given in Table 1. The prediction will improve if we add hysterical index factor to streamflow prediction equation.

Conflicts of Interest. The authors declare no conflict of interest.

References

1. Davies JAC, Beven K (2015) Hysteresis and scale in catchment storage, flow and transport. Hydrol Process 29:3604–3615. https://doi.org/10.1002/hyp.10511
2. Beven K (2006) A manifesto for the equifinality thesis. J Hydrol 320(1–2):18–36
3. Sproles EA, Leibowitz SG, Reager JT, Wigington PJ, Famiglietti JS, Patil SD (2015) GRACE storage-runoff hystereses reveal the dynamics of regional watersheds. Hydrol Earth Syst Sci 19:3253–3272. https://doi.org/10.5194/hess-19-3253-2015
4. Aich V, Zimmermann A, Elsenbeer H (2014) Quantification and interpretation of suspended-sediment discharge hysteresis patterns: How much data do we need? Catena 122:120–129. https://doi.org/10.1016/j.catena.2014.06.020
5. Klein M (1984) Anti clockwise hysteresis in suspended sediment concentration during individual storms: Holbeck catchment; Yorkshire, England. Catena 11:251–257. https://doi.org/10.1016/0341-8162(84)90014-6
6. Seeger M, Errea MP, Beguería S, Arnáez J, Martí C, García-Ruiz JM (2004) Catchment soil moisture and rainfall characteristics as determinant factors for discharge/suspended sediment hysteretic loops in a small headwater catchment in the Spanish pyrenees. J Hydrol 288:299–311. https://doi.org/10.1016/j.jhydrol.2003.10.012
7. Ang WT, Garmón FA, Khosla PK, Riviere CN (2003, October) Modeling rate-dependent hysteresis in piezoelectric actuators. In: Proceedings 2003 IEEE/RSJ International Conference on Intelligent Robots and Systems (IROS 2003) (Cat. No. 03CH37453) (vol 2, pp 1975–1980). IEEE
8. O'Kane JP (2005) Hysteresis in hydrology. Acta Geophys Pol 53(4):373–383
9. Lawler DM, Petts GE, Foster ID, Harper S (2006) Turbidity dynamics during spring storm events in an urban headwater river system: The Upper Tame, West Midlands, UK. Sci Total Environ 360(1–3):109–126
10. Malhi Y, Roberts JT, Betts RA, Killeen TJ, Li W, Nobre CA (2008) Climate change, deforestation, and the fate of the Amazon. Science 319(5860):169–172
11. Phillips CL, Nickerson N, Risk D, Bond BJ (2011) Interpreting diel hysteresis between soil respiration and temperature. Global Change Biol 17(1):515–527
12. Almeida GAM, Bates P, Freer JE, Souvignet M (2012) Improving the stability of a simple formulation of the shallow water equations for 2-D flood modelling. Water Resour Res 48:W05528. https://doi.org/10.1029/2011WR011570
13. Bunn SE, Arthington AH (2002) Basic principles and ecological consequences of altered flow regimes for aquatic biodiversity. Environ Manage 30(4):492–507

14. Lytle DA, Poff NL (2004) Adaptation to natural flow regimes. Trends Ecol Evol 19(2):94–100
15. Poff NL, Zimmerman JK (2010) Ecological responses to altered flow regimes: a literature review to inform the science and management of environmental flows. Freshw Biol 55(1):194–205
16. Richter B, Baumgartner J, Wigington R, Braun D (1997) How much water does a river need? Freshw Biol 37(1):231–249
17. Baker DB, Richards RP, Loftus TT, Kramer JW (2004) A new flashiness index: characteristics and applications to midwestern rivers and streams 1. JAWRA J Am Water Res Asso 40(2):503–522
18. Butturini A, Guarch A, Romaní AM, Freixa A, Amalfitano S, Fazi S, Ejarque E (2016) Hydrological conditions control in situ DOM retention and release along a Mediterranean river. Water Res 99:33–45

Chapter 16
Scouring Due to a Vertical Jet: A Review of Parameters

Lalit Yadav[✉] and Baldev Setia

Department of Civil Engineering, NIT Kurukshetra, Kurukshetra, Haryana 136119, India

Abstract. Hydraulics engineers have always shown keen interest in the investigation of scouring as a phenomenon, which can be attributed to the role that scouring plays in deciding the safety and efficiency of the hydraulic structures. The presence of vertical jets in real life can be felt in ski jump spillways, hydraulic jump-type stilling basins, vertical gates, and the cleaning of reservoirs for sediments. In the present study, an attempt is made to present a detailed review of the up-to-date work on scouring due to a vertical jet. It includes the variation in the scour depth under the dynamic and static conditions of the submerged and non-submerged state, factors responsible for the effect produced under a jet, and a comparison of parameters used by various researchers. A lot of work has been conducted to assess the effect of various functional parameters, namely the type, size and diameter of sediment, impinging height, velocity, and diameter of the jet, etc., on scour characteristics. Relatively lesser attention has been given to the impinging angle of the jet and the shape of the nozzle of jet. The inferences drawn from the review of the quantum of work done on the scouring due to a vertical jet have proved helpful in the understanding of the phenomenon of scour.

Keywords: Hydraulics · Vertical jet · Dynamic scour · Impinging angle

List of Symbols

E_c	Erosion parameter for circular jets
D	Median size of sediment
T	Experimental run-time
T_m	Mean average value of T
V	Velocity of the falling jet
H	Impinging height of the jet
H'	Depth of scouring
d	Nozzle diameter of the jet
y_m	Maximum or equilibrium scour depth
Fo	Densimetric Froude number
g	Acceleration due to gravity
ρ	Mass density of fluid
$\Delta\rho$	Difference between mass densities of bed material and fluid

R. K. Bhattacharjya et al. (eds.), *Sustainable Water Resources Management*, Advances in Sustainability Science and Technology,
https://doi.org/10.1007/978-981-16-7535-5_16

1 Introduction

Scour is a term that is used to report the noteworthy localized erosion of earth material, occurring when the ability of earthen material to withstand the erosive capacity of water becomes much less than the displacing power of the water [20]. Scouring due to turbulent nature of water may be visualized as a continuous recurring phenomenon; i.e. it may be said to be responsible for the cutting off of large chunks of soil and sediment from certain specific locations of hydraulic structures including from around bridge piers, near the abutments of bridges, in case of plunge pools, and to be occurring downstream of a dam. Hydraulic structure demand is on the rise so as to fulfil the needs for drinking water, irrigation, flood control, recharging of groundwater resources, and generation of electricity, for both domestic and industrial requirements. Flows through hydraulic structures often issue in the form of jets. The velocities of these jets are usually high and high enough for them to result in considerable and dangerous scour holes, which might impact the safety of the structure. In the case of a dam, the surplus water contained in the reservoir is discharged into the spillways, and then from spillways, this water is issued in the form of jets to the stilling basins. If the stilling basins are absent on the downstream of a dam, then scouring is bound to take place due to the impact of impinging jet on the river bed downstream. Various studies have been performed for investigating the erosion of cohesive and non-cohesive materials under the submerged and non-submerged conditions of the impinging jet. The importance of these studies has already been highlighted above. Some of the pioneering works carried out in this area include those of Doddiah et al. [5], Westrich and Kobus [22], Rajaratnam and Beltaos [17], Mih and Kabir [14], and Aderibigbe and Rajaratnam [1].

2 Literature Review

As available in the open literature, the first experiments on scour due to a submerged vertical jet were conducted by Rouse [21]. It was followed by other investigators who worked on this thrust area, and one such work was by Doddiah et al. [5]. They are regarded as pioneers in this field, as the authors had not only worked on hollow and solid jets but also compared their study with that of scour at the base of a free overfall. Results obtained showed that the scour due to a hollow jet was approximately the same as that caused by the solid jet. Further, the depth of scour was found to be dependent on time. Also, the scour depth was found to be more predominantly affected by the increase in discharge as compared to an increase in drop height.

To estimate the relative scour resistance of remoulded and natural sediments, Moore and Masch [15] conducted a series of tests using a vertical submerged round impinging water jet. The rates of scouring were calculated by considering the loss of weight in the sample. The mean depth of scour was found to be proportional to the logarithm of the time during which scour occurred. Westrich and Kobus [22] and Kobus et al. [9] conducted experiments using variable mean velocity, two types of nozzle diameter, and two sets of jet height for studying the phenomenon of jet scour on a uniform sand bed with vertically submerged jets. Rajaratnam [16] made use of the plane wall and impinging jets of air and water on the beds of sand and polystyrene in order to study the erosion caused. Well-designed nozzles of thicknesses 2.41 and 2.5 mm were employed for producing air jets and water jets, respectively. From the results obtained for both the sets of experiments, one common inference deduced was that the dynamic scour depth was found to be larger than static scour depth. He also proposed that maximum depth of erosion (y_m) is a function of densimetric Froude number (Fo).

$$\text{Fo} = V/\sqrt{g(\Delta\rho/\rho)D}$$

Aderibigbe and Rajaratnam [1] conducted experiments in an octagonal box with dimensions 0.235 × 0.6 m (side length × height) to observe the maximum depth of scour as a result of variation in the impinging distances of jet from the sand bed. Other experimental parameters like jet velocity, diameter of the nozzle, and the median size of the sediments used were kept constant. They developed empirical relations for the maximum scour depth and scour hole radius and defined and proposed an expression for erosion parameter (E_c), further relating it to the characteristic length of eroded bed. The authors suggested the presence of two flow regimes, i.e. a strongly deflected jet regime (SDJR) and a weakly deflected jet regime (WDJR).

$$E_c = V(d/H)/\sqrt{(gD\Delta\rho/\rho)}$$

Mazurek et al. [11] made use of cohesive soil for studying the scour produced by a submerged vertical jet. Impingement height of the jet was varied from 40 to 116 mm, and the jet nozzle was submerged up to 100 mm. The study suggested three types of erosion, i.e. flake erosion, erosion of small or large chunks, and rapid surface erosion. Based on the mechanics of the impinging jets, the study revealed dependence of scour hole on the momentum flux and impingement height of the jet, viscosity and density of the fluid, and the critical shear stress of the soil. A combination of cohesionless and cohesive sediments was used by Ansari et al. [2] for studying the scour under submerged vertical jets of water and further for the identification of variation in scour patterns. On comparison, the maximum scour depth in case of non-plastic sediment was found to be similar to the cohesionless sediment. Also, empirical relationships for the scour volume and maximum depth of scour for both plastic and non-plastic sediments were proposed by the authors. A high vertical drop was used by Dey and Raikar [4] for studying the scour in case of sand and gravels. For the analysis, variation in the jet thickness and the depth of tailwater was incorporated. The authors also established a relationship between the equilibrium scour depth (y_m) and densimetric Froude number (Fo). Observations also suggested the dependence of the equilibrium scour depth on sediment size and tailwater depth. Other researchers to consider the use of freely falling jets for studying the scour were Ghodsian et al. [6]. The parameters used in their study were densimetric Froude number (Fo), tailwater depth, drop height, and the sediment size. Results obtained from the study indicate that there is a decrease in the scour hole parameters with an increase in the sediment non-uniformity. Maximum scour depth, ridge height, length of scour, and the width of scour hole were found to be dependent on the drop height. The authors also suggested the use of d_{90} instead of d_{50} for the sediments used.

Chakravarti et al. [3] observed the variation between dynamic scour depth and static scour depth in the case of submerged circular vertical jets. The authors utilized four parameters, namely velocity of jet, impinging height of the jet, nozzle diameter, and the size of sediment for the purpose of studying the radius, volume, and the maximum scour depth (y_m). The study established the variation of maximum scour depth (y_m) with the erosion parameter (E_c) under submerged circular vertical jet in cohesionless sediments. The effect of air entrainment was considered by Maleki and Fiorotto [10] in studying the local scour depth and shape, downstream of rectangular falling jets for high hydraulic structures. An attempt was made by the authors to correlate the theoretical analysis with that of the laboratorial evidence. For exploring the role of the sediment non-uniformity in attaining the equilibrium scour depth, experimental runs with sediments having $d_{50} =$ 8 and 14 mm and specific discharge rates equal to 0.03 and 0.04 $m^3/s/m$ were performed. The authors observed the presence of scouring in both the wall jet and the impingement region. Further, the impingement region scouring was found to be dependent on the shear stress and the bottom pressure changes. It was established that the scour depth is dependent on the densimetric Froude number (Fo), angle of repose, and the lift coefficient of sediments.

Table 1 presents a summary of the works and range of parameters of some of the significant researchers.

From a review of the available literature on the subject, it is amply clear that it has been an interesting subject of research drawing the attention of various investigators during the last few decades. The occurrence of similar phenomena in the field has been another reason for research in this thrust area to have sustained during all this time. It has been seen that though the variables are not many yet the range of these variables used by the previous researchers has been wide. Some of the researchers have used some subjective terminology like 'until equilibrium stage' but have not described any objective criteria to judge this stage.

The following article makes an attempt to have a deeper insight into the range of these variables which will help in drawing some inferences regarding the scaling up of the results of experiment, if required.

3 Comparison of Experimental Parameters Used by Different Researchers

Scouring due to a vertical jet has been investigated by a number of researchers working under both the submerged and non-submerged conditions of the jet. Various parameters like velocity of falling jet, impinging height of jet, nozzle diameter of jet, sediment size, experimental run-time, and the impinging angle of the jet have been used to study the scour hole profile for both the cohesive and the non-cohesive sediments. However, there is a significant variation in the parameters adopted by the researchers.

Experimental Run-Time, '*T*'

The experimental run-time used by various researchers ranges over a very wide base. On the higher side, it includes the experimental run-time adopted by Rajaratnam [16] (66 h), Mazurek et al. [11] (24 h), **and** Ghodsian et al. [6] (64 h), and on the other side, the minimum run-time being employed by Hou et al. [7] is equal to barely 1/12 h (5 min). On simple comparison, it is difficult to assess as to why the researchers of the two categories

Table 1 Summary of the works of various researchers

Author	Notation for authors	Type of work	Velocity, V (m/s)	Median size, D (mm)	Impinging height of jet, H (m)	Nozzle diameter, d (mm)	Time, T (h)
Doddiah et al. [5]	D, M, and R	Circular, free overfall jet	0.2926, 0.2194; 0.365	0.3556	0.7112, -	0.3048, –	Until equilibrium scour
Rajaratnam [16]	N	Wall jets of air and water	–	1.2	0.5	2.41, 2.55	66
Aderibigbe and Rajaratnam [1]	O and N	Submerged vertical jets	2.65–4.45	0.88, 2.42	0.004–0.523	4,8,12,19	6–50
Mazurek et al. [11]	K, N, and S	Vertical jet on cohesive soil	4.97–25.9	–	0.004–0.116	4, 8	24
Rajaratnam and Mazurek [18]	N and K	Circular jets with small tailwater	1.60–4.22	1.0, 1.15, 2.38	0.048–0.318	9.8, 12.7	24
Mazurek et al. [12]	K, N, and D	Plane turbulent wall jets	4.86–13.56	–	–	2.33, 5.10	24
Ansari et al. [2]	S, U, and R.R	Influence of cohesion on scour	1.3–5.75	0.27	0.15, 0.30	8.0, 12.5	Until equilibrium scour

(*continued*)

Table 1 (*continued*)

Author	Notation for authors	Type of work	Velocity, V (m/s)	Median size, D (mm)	Impinging height of jet, H (m)	Nozzle diameter, d (mm)	Time, T (h)
Rajaratnam and Mazurek [19]	N and K.A	Circular jets on rough boundaries	45–90	0.288, 1.37,2.53	0.15–0.33	6.4–12.7	–
Dey and Raikar [4]	S and R	High vertical drop	–	0.26,0.49,0.81, 1.86,2.54,4.1, 5.53,7.15	4.3–29.8 (relative)	–	9, 12
Ghodsian et al. [6]	M, M, and R	Free fall jets, non-uniform sediments	–	1.28	0.6647, 0.771, 0.821, 0.871	–	64
Mehraein et al. [13]	M, M, and A	Simultaneous wall **and** impinging jets	–	0.6, 1.05	–	20, 10	24, 52
Chakravarti et al. [3]	A, R, and U	Circular vertical jet	5.12–9.84	2.8	0.15–0.30	8–12.5	Until equilibrium scour
Hou et al. [7]	J and Z	Submerged water jet	–	0.05	0.03–0.084	3	1/12
Maleki and Fiorotto [10]	S and V	Falling plane jet		3, 5, 9	0.15 to 0.4	–	18

would have justified these extreme experimental conditions. Figure 1 presents the time durations adopted by different authors for conducting their experiments. In the figure, the authors have been sorted in accordance with increasing time and listed along the x-axis and the run-time has been given along the y-axis and shown with the help of bar charts.

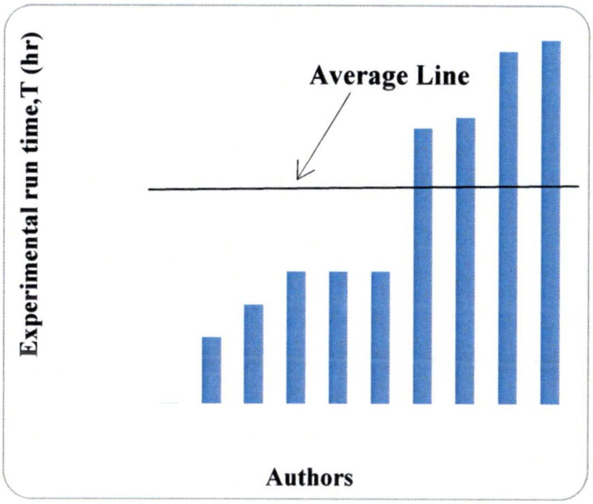

Fig. 1 Variation in the experimental run-time

From Fig. 1, the wide range of experimental run-time is obvious. The arithmetic mean value of the experimental run-time 'T_m' used by the researchers works out to be 33.4 h. Thus, 40% of the researchers mentioned in the paper have worked above and 60% of researchers have worked below the mean value of experimental run-time. Interestingly, majority of the authors have not offered any comments on the logic of choosing that particular duration of time. Three of the papers have chosen a subjective statement, i.e. 'until equilibrium scour' to describe the time duration, but have not specified any criteria to decide that stage.

Velocity of the Falling Jet, 'V'

Velocity of the falling jet is one of the prime parameters considered for the study of scouring due to a vertical jet in both cohesive and non-cohesive sediments. It is a measure of the impact of the jet and hence directly affects the scour and scour hole characteristics. Researchers have worked upon a range of velocities to study the extent of scouring in terms of depth, radius, and volume. For the ease of understanding, an average of the range of velocities is taken and represented in Fig. 2 with a horizontal line.

Figure 2 shows that the average value for the varying jet velocities used by the researchers comes out to be 10.89 m/s which indicates that 20% of the researchers have worked with a jet velocity greater than the average value while the remaining 80% of the researchers have worked below the mean average value of velocity of the falling

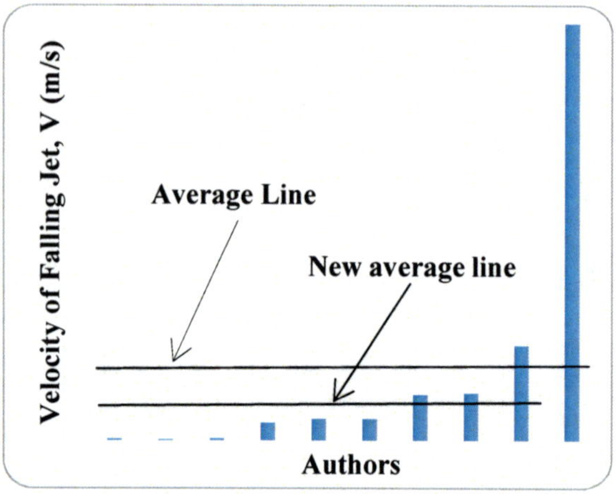

Fig. 2 Variation in the velocity of falling jet

jet. Further, that for nine out of 10 lower values, the velocities have risen gradually, while for the ultimate maximum value, the velocity is far too high to be a part of the general trend. This value may safely be excluded from the general population, and a new average 4.609 m/s may be proposed, represented by lower horizontal line in Fig. 2. The figure also manifests that there is no particular criteria under which the velocity has been decided by different authors.

Sediment Size, '*D*'

The effect of variation in sediment size predominantly affects the process of scouring irrespective of the type of jets, i.e. either wall jets or impinging jets. Further, this fact can be established from the literature as the researchers have previously worked upon both the cohesive and non-cohesive sediments. In the case of non-cohesive sediments, the scour is an immediate response to the magnitude of total energy impacting the sediment of a particular d_{50}. In Fig. 3, a horizontal line demarcates sediment size above and below the average d_{50} value.

Figure 3 shows that 25% of the researchers have worked with the sediment size more than the average value of 1.65 mm (d_{50}) found using the simple arithmetic mean while 75% of the researchers have worked with the sediment size less than the average value. In fact, the largest sediment size falls in the category of gravel and that single size is responsible for a large mean value. Also, the mean value is larger than the D50 value of River Yamuna at Yamuna Nagar ($d_{50} = 0.32$ mm) and River Ganga at Kanpur ($d_{50} = 0.16$ mm). When preliminarily compared with the curve on velocity (Fig. 2), no particular correlation between the two parameters can be established.

Impinging Height of the Jet, '*H*'

The vertical distance between the nozzle of the jet and the surface of the sediment bed depicts the impinging height of the jet. Figure 4 shows various values of impinging heights used by the researchers for determining the scour in case of impact of a jet.

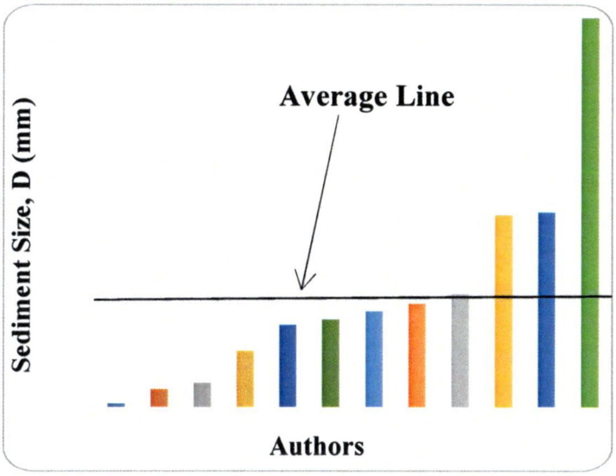

Fig. 3 Variation in the sediment size, *D*

From Fig. 4, the average value for the varying impinging heights comes out to be 1.52 m. On observation, it is found that 93% of the researchers have worked with an impinging height below the average value while 7% of the researchers have worked above the average value. Here too, barring one value, the others are falling in a band of low diversity. Therefore, with the largest value as the outlier, the average of the remaining data comes out to be 0.415 m/s, which has been depicted by the lower horizontal line in Fig. 4.

Nozzle Diameter, '*d*'

Nozzle diameter depends upon type of sediment (cohesive or non-cohesive) and the type of jet (wall jet or impinging jet). Figure 5 shows the various nozzle sizes used by the researchers. The maximum size of the nozzle diameter used is 20 mm, while the minimum is 0.3048 mm. The average value of the nozzle diameters used by the researchers considered in this study is found to be 8.27 mm indicating that 41% of the researchers have worked above and 59% below the average line. Perhaps, this is a parameter where the values seem to be fairly distributed on both sides of the average line.

4 Conclusion

Broadly, it may be seen and appreciated that there is a large variation in the use of different parameters adopted for experimentation making it difficult to generalize the results obtained through such experiments. Despite a significant quantum of work done on the subject of scouring due to a vertical jet, a relatively much lesser attention has been given to the parameters like angle of jet and shape of nozzle. Major conclusions of the study are:

- In case of experimental run-time, 40% of the researchers considered in the present paper have used time duration of more than the T_m value.

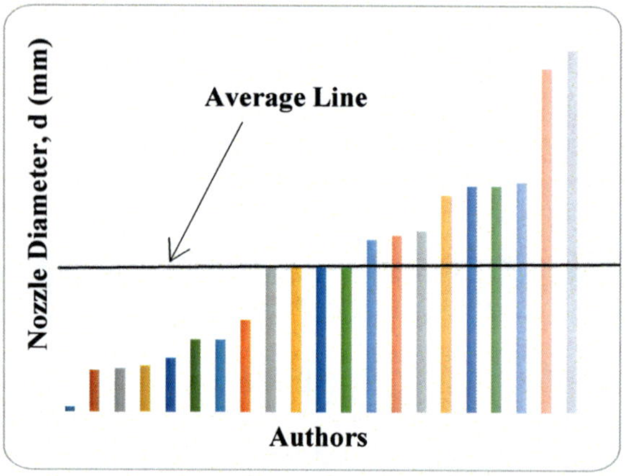

Fig. 4 Variation in the impinging height of the jet

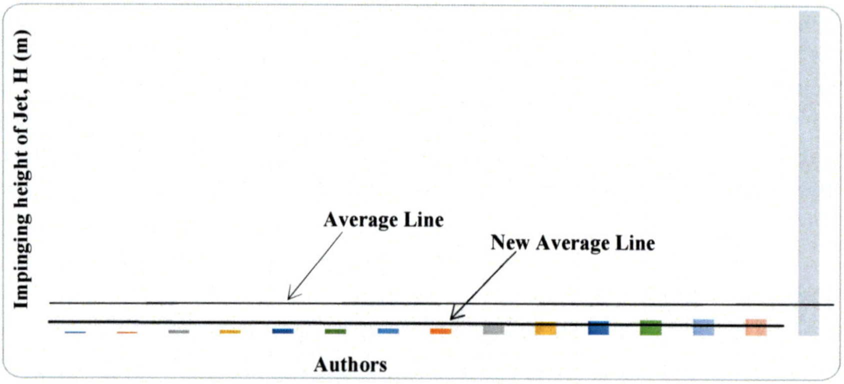

Fig. 5 Variation in the nozzle diameter

- Average value of velocity of the falling jet in present collection is 10.89 m/s, suggesting that 20% of the researchers have worked above and 80% below the average line.
- In case of the sediment size, 25% of researchers had worked above and 75% of the researchers worked below the average sediment size, d_{50} value.
- In case of the impinging height of the jet, 7% of researchers had been working above and 93% researchers below the average value of the impinging height of the jet.
- When the average value of the nozzle diameter of jet is compared, results showed that 41% of the researchers had worked above and 59% below the average line.

5 Way Forward

Though a significant quantum of work has been done on the subject of scouring due to a vertical jet, there is a large variation in the use of different parameters. In majority of studies, the choice of a particular range of parameters has also not been supported by any criteria. The field of investigation is still wide open, both theoretically and experimentally. The following areas are still grey and need exploration.

(a) Relatively much lesser attention has been given to the parameters like angle of jet and shape of nozzle.
(b) The combined effects of wall jets and vertical jets must be incorporated in the future studies. It is also observed that most studies have been done around a single parameter, i.e. erosion parameter, while other parameters like specific energy should also be given due preference for studying the impact of jet on the sediments.
(c) Under dynamic conditions, the initial strike of jet on the sediment is considered as the impinging height of jet (H), while the scoured depth (H') should also be added in the analysis.
(d) Mechanism behind the formation and expansion of jet has altogether been missing in the previous studies and should be given due consideration in the future studies.
(e) Efforts should also be made towards suggesting and recommending guidelines for a good experimental set-up.

Acknowledgements. First, the author would like to convey grateful thanks to Mr. Rahul Malik (PhD Scholar) for his constant support and guidance all through the preparation of the manuscript.

References

1. Aderibigbe OO, Rajaratnam N (1996) Erosion of loose beds by submerged circular impinging vertical turbulent jets. J Hydraul Res 34(1):19–33
2. Ansari SA, Kothyari UC, Ranga Raju KG (2003) Influence of cohesion on scour under submerged circular vertical jet. J Hydraul Eng 129(12):1014–1019
3. Chakravarti A, Jain RK, Kothyari UC (2014) Scour under submerged circular vertical jets in cohesionless sediments. ISH J Hydraul Eng
4. Dey S, Raikar RV (2007) Scour below a high vertical drop. J Hydraul Eng 133(5):564–568
5. Doddiah D, Albertson ML, Thomas R (1953) Scour from jets. In: Proceedings of 5th international association for hydraulic research congress, IAHR, Minneapoli, USA
6. Ghodsian M, Mehraein M, Ranjbar HR (2012) Local scour due to free fall jets in non-uniform sediment. Scientia Iranica A (2012) 19(6):1437–1444
7. Hou J, Zhang L, Gong Y, Ning D, Zhang Z (2016) Theoretical and experimental study of scour depth by submerged water jet. Adv Mech Eng 8(12):1–9
8. Jain RK, Kothyari UC (2010) Influence of cohesion on suspended load transport of non-uniform sediments. J Hydraul Res 48(1):33–43
9. Kobus H, Leister P, Westrich B (1979) Flow field and scouring effects of steady and pulsating jets on a movable bed. J Hydraul Res 17(2):175–192
10. Maleki S, Fiorotto V (2019) Scour due to a falling plane jet: a comprehensive approach. J Hydraul Eng 145(4):04019008

11. Mazurek KA, Rajaratnam N, Sego DC (2001) Scour of cohesive soil by submerged circular turbulent impinging jets. J Hydraul Eng 127(7):598–606
12. Mazurek KA, Rajaratnam N, Sego DC (2003) Scour of a cohesive soil by submerged plane turbulent wall jets. J Hydraul Res 91(2):195–206
13. Mehraein M, Ghodsian M, Schleiss AJ (2012) Scour formation due to simultaneous circular impinging jet and wall jet. J Hydraulic Res 50(4):395-399
14. Mih WC, Kabir J (1983) Impingement of water jets on non uniform streambed. J Hydraul Eng ASCE 109(4):536–548
15. Moore WL, Masch FD (1962) Experiments on the scour resistance of cohesive sediments. J Geophys Res 67(4):1437–1446–1446
16. Rajaratnam N (1981) Erosion by submerged circular jets. J Hydraul Div ASCE 108(HY2):262–267
17. Rajaratnam N, Beltaos S (1977) Erosion by impinging circular turbulent jets. J Hydraul Div ASCE 103(10):1191–1205
18. Rajaratnam N, Mazurek KA (2002) Erosion of sand by circular impinging water jets with small tail water. J Hydraul Eng ASCE 129(3):225–229
19. Rajaratnam N, Mazurek KA (2005) Impingement of circular turbulent jets on rough boundaries. J Hydraul Res IAHR 43(6):688–694
20. Raudikivi AJ (1992) Loose boundary hydraulics, chap. 9, 3rd edn. Pergamon Press, New York
21. Rouse H (1939) Criteria for similarity in the transportation of sediment. Bulletin 20, University of Iowa, Iowa, USA, pp 33–49
22. Westrich B, Kobus H (1973) Erosion of a uniform sand bed by continuous and pulsating jets. Proc., 15th IAHR Congress, Istanbul, Turkey, vol 1, A13.1–A13.8

Chapter 17
Identification and Mapping of 2019 Flood Extents Using Sentinel-1 A Images: A Case of Barpeta District, Assam

Leena Chetia(✉), Saikat Kumar Paul, Richa Dhawale, and Nayana Merin Joy

Department of Architecture and Regional Planning, Indian Institute of Technology Kharagpur, Kharagpur, West Bengal, India
skpaul@arp.iitkgp.ac.in

Abstract. During the time of a crisis, the foremost initiation is decision making and minimizing its extent, for which, real-time crucial information is required. In flood scenarios, it is always challenging to get quick, efficient, and reliable information due to severe weather conditions and disruption in all means of communication. Even though optical radar image reveals various earth surface information, it has limitations during severe weather conditions, whereas in the case of synthetic aperture radar (SAR), the emitted microwaves can acquire remotely sensed high-resolution real-time spatial information, in spite of the fog, mist, or cloudy conditions during monsoon season. The SAR sensors are highly sensitive to the roughness and wetness of the surfaces; and as it is an active satellite system, it can acquire ground information at any hour of the day. The near real-time SAR information can help in swift mapping of flood situations helping in decision making, responding, and minimizing massive loss of life and assets. Therefore, SAR data has been preferred for the study. In this paper, the potential of Sentinel-1 A, high-resolution C band image, along with GIS, is explored, taking the ground scenario of the 2019 Assam flood into consideration. A series of SAR crisis images (during the flood) of four days, that is 14th, 21st, 26th July and 7th August 2019, and one archived image (non-flooded time), 5th December 2019, were selected as candidate images. They were processed and analysed, using ESA SNAP Sentinel-1 Toolbox and GIS. The spatial extent of flood and inundation of built-up areas have been estimated for this period. The study found that the variation was approximately 219 km^2 in archive time to 1089 km^2 on the extreme situation that is on 26th July 2019. The flood situation improved from 14th to 21st and again intensified on 26th July. Gradually towards 7th August, the condition was improved, affecting only about 78 km^2 of built-up area. The findings of the study can help in understanding the miserable situations and flood extents, decision making, minimize the impact of floods, and lead to risk reduction and better management during disastrous situations.

Keywords: Remote sensing · SAR data · Sentinel-1 · ESA SNAP · Flood mapping · Disaster response

R. K. Bhattacharjya et al. (eds.), *Sustainable Water Resources Management*, Advances in Sustainability Science and Technology,
https://doi.org/10.1007/978-981-16-7535-5_17

1 Introduction

Flood is a situation of inundation of areas that normally remains dry. Amongst all types of natural disasters, it is noticed that flood is most common and most destructive in nature [6, 13], believed to be contributed mostly by global climate change, which is the prime threat to mankind in twenty-first century [10]. Considering all types of natural disasters, flood contributes to 23% of total economic loss due to natural disasters, 43.4% of total disasters of different types, affecting 45% of total affected, and death of 11% of people [14]. Globally, India positions itself amongst the top ten countries/territories in terms of absolute losses (billion US$) from 1998 to 2017, and 79.5% loss is due to flood [9, 14]. Annual flood conditions have always threatened the life and livelihood in Assam, India.

The recent incidents of catastrophic floods and resulting losses have demanded for prompt measures against it [4, 15]. To assess the flood situations, the major challenge is to gain real-time or near real-time flood data [2, 11]. The conventional methods require manpower, training, and extensive surveys which is time-consuming, expensive, and also prone to human errors [1, 12] Furthermore, due to the severity of floods and loss of communication makes it impossible to achieve data from the site during crisis time [7]. On the contrary, the various active and passive satellite missions on earth observation have made easy availability of space-born earth surface images [4, 5, 11]. SAR is such an active satellite system. The European Space Agency (ESA) Sentinel-1 provides options of S1A and S1B with low to high-resolution images with different bands which reveals the detailed and near real-time ground information, which can be used for investigation of flood conditions [11, 16]. These data can be used for effective information extraction and utilization for disaster management, for example: identification and monitoring of flood extents in an affected area [3, 8]. This study identifies the spatial extent of the recent flood event (14th July to 7th August 2019) and the inundation status of built-up areas in Barpeta District, Assam. The study is carried out by using SAR data, Sentinel-1A, C band, in ESA STEP SNAP Sentinel-1 Toolbox, and the final analysis is done on Arc GIS. The results intend to help in better flood management and disaster risk reduction.

2 Geographic Setting of the Study Area

Due to frequent and prolong flood effects, the Barpeta District of Assam is selected for the study. It is located on the lower division of Assam. It is situated between 90°39′30″–91°23′00″E longitude and 26°05′30″–26°48′30″N latitude (Fig. 1). It has a total land area of 2282 km^2, containing nine revenue circles (RC). As per the 2011 census, the population of the district is 1,693,622 with a density of 520/km^2. It is surrounded by Nalbari District towards East, Bongaigaon and Chirang District towards West, Baksa District and Bhutan towards North and river Brahmaputra, and Kamrup District towards South. The major rivers in the area are Beki, Manah, Pohumara, Kaldia, Palla, Nakhanda, Marachaulkhowa, and Bhelengi. Pohumara and Kaldia join and form river Nakhanda; on the other hand, Palla and Beki join with Nakhanda to ultimately form Chaulkhowa River. Every year the heavy rainfalls in the area during monsoon season, low elevation, the mighty Brahmaputra River towards its complete Southern stretch, and complex network of rivers make the district highly flood prone. In the 2019 Assam flood, 8 RCs out of the 9 were affected.

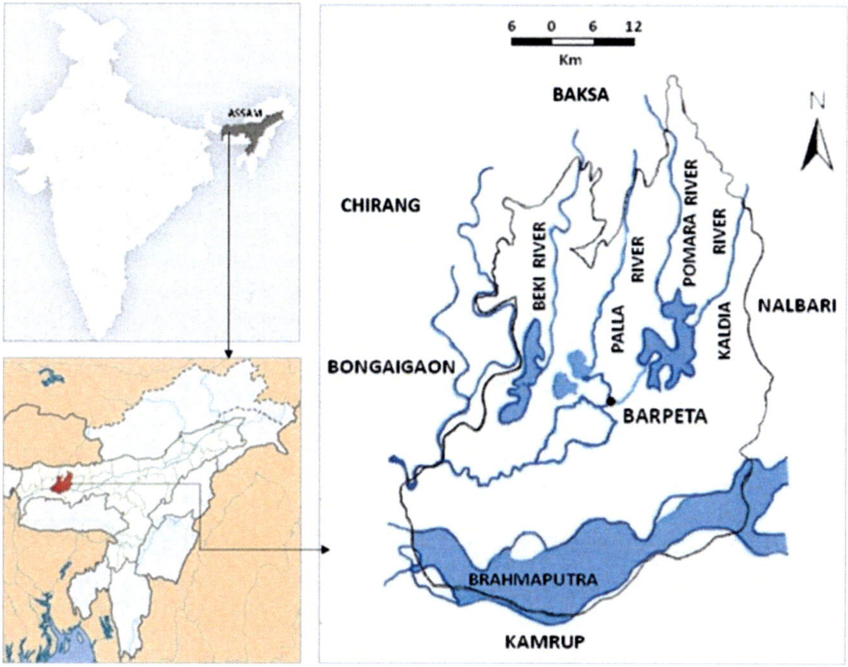

Fig. 1 Location of Barpeta District, Assam, India

3 Data Used and Methodology

Data Used

As SAR data can provide reliable earth surface information even in severe weather conditions, it was preferred for the study. The Sentinel-1 A, C band, Interferometric Wide (IW) swath mode, standard class, Level-1 Ground Range Detected (GRD), high-resolution (H) data with dual polarization (VH, VV), and 10 m pixel spacing was used. The data of 14th, 21st, 26th July and 7th August 2019 that is the crisis time was used for flood identification and delineation of flood extent. Similarly, to extract the open water areas and compare the flood extents, an image of the non-flooded time of 5th December 2019 was used. On the other hand, Sentinel-1 A, C band IW, simple look complex (SLC) data of 8th and 20th December 2019 was used to extract the built area footprints.

Methodology

The brief methodology to carry out the study is given in Fig. 2. The process was carried out in two broad stages: (i) Image processing—the raster-based analysis was carried out in ESA STEP SNAP Sentinel-1 Toolbox, and (ii) the GIS-based analysis was carried out in ArcGIS.

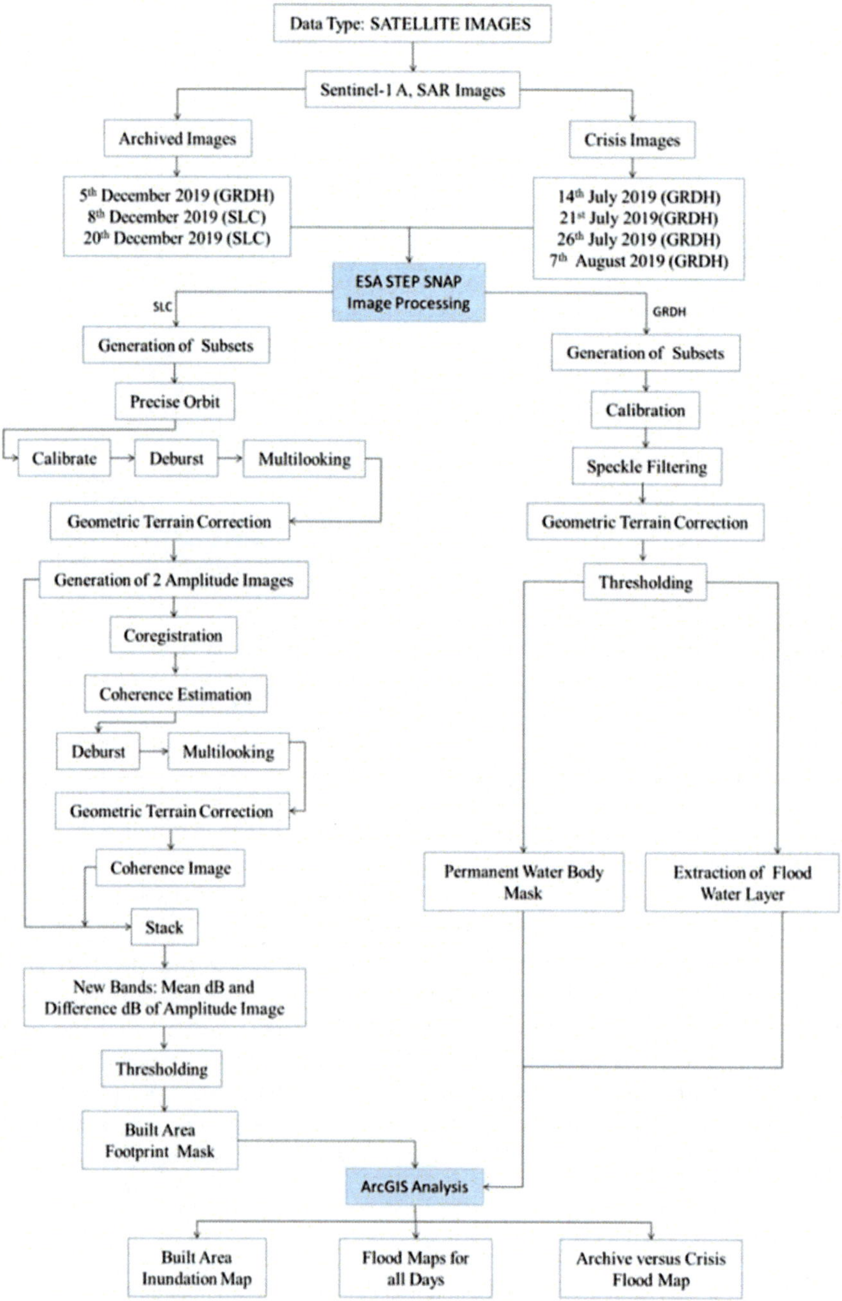

Fig. 2 Brief methodology

Preprocessing of SAR Images

The images require preprocessing to carry out further analysis, which was done in a sequential manner using ESA SNAP Sentinel-1 Toolbox. First of all, the image subsets were created for the interested area. It was then calibrated using radiometric calibration in order to transform simple digital pixel values to physical quantities in (the backscattering) sigma naught images. As the differences are very high in the images, they were then converted to decibel (dB) formats, and the bands were rewritten. The SAR images are not terrain corrected and geocoded; also, a physical phenomenon called "speckles" is generated due to the interaction of radar pulses with different scatterings from the earth's surface, and it is generally corrected by using filtering techniques. The calibrated images were filtered with "Lee Filter" 3 X 3 window, to remove the speckles and then converted to dB format. For the filtered images, the terrains were referenced to SRTM 1Arc Sec Digital Elevation Model (DEM) and were geocoded to Geographic Coordinate System-WGS1984. The image bands were again converted to dB format and image bands were rewritten.

Similarly, to extract the footprints of built areas, two archived Sentinel-1 A, IW, SLC, 1SDV images of 8th and 20th December 2019 were taken as candidate images. The amplitude and phase information is used. The amplitude and the difference of interferometric coherence between the two images are calculated. The selected bands for the subset were assigned to precise orbit, calibrated, multilooked, geometrically terrain corrected, and the two amplitude images were generated and transferred to dB format. These two amplitude images were then co-registered, and coherence was estimated and further debursted, multilooked, and geographically terrain corrected to generate the final coherence image. The bands of coherence images were again converted to dB format. The two amplitude images and the coherence image were again staked together. New bands, the mean dB and difference dB was generated using the amplitude images and was written to the image. The mean and difference dB and the coherence image were used to mask out the built area footprints.

The calibrated, speckle filtered, and geometrically corrected images were used for the generation of flood maps and analysis in ArcGIS.

Generation of Flood Maps

To identify the flooded areas, information on two major classes is required—the permanent water bodies, for example: river channels, lakes, ponds, wetlands, etc. and the land areas. The range of water layer values in SAR images varies between -15 and -25 dB [1]. The permanent water bodies were extracted from the archived image (5th December 2019), using the thresholding method. After multiple thresholding trials, -19 dB was set as the limit. The condition was given: land, if dB > -19 and water if dB ≤ -19, and permanent water bodies were unmasked.

With respect to the permanent water layer, the preprocessed SAR images for respective dates were analysed in ArcGIS. The crisis images were overlapped, and a flood water layer for each crisis day was generated. The flood maps were then analysed to understand the growth of the flood in this duration.

Identification of Inundation of Built Areas

To identify the inundation of total built areas in the district, the built area footprints were unmasked using ESA SNAP Sentinel-1 Toolbox, and threshold conditions were given as "if mead dB > -10, and coherence (of 20th Dec. image) is > 0.6 then 1 else 0". The processed image was further analysed in ArcGIS by overlapping with the extreme flood situation, and the inundated built footprints were identified using the intersection tool in Arc GIS.

4 Results and Discussion

The flood situations of the considered timeline are shown in Fig. 3. From the analysis of the SAR images, the spatial extent of flood in the Barpeta District, Assam, was identified and the total inundation along with the built area was estimated. The open water layer area was estimated to be approximately 219 km^2, which is 9.5% of the total land area (Fig. 3a). During flood situation, it was observed that on 14th July 2019, the inundation was approximately 996 km^2 of land (Fig. 3b), whereas on 21st July, it was gradually reduced to 972 km^2 (Fig. 3c), and by 26th of July, the flood extent was raised to maximum by occupying 1089 km^2 (Fig. 3d), which is approximately 47.7% land area of the district. On the contrary, it was decreased to 793 km^2 on 7th August 2019 (Fig. 3e). So, it can be said that the severity of flood from 14th to 21st j July improving but again raised on 26th July making it to be an extreme flooded day, and after that, it receded by 7th August 2019. The extreme flood situation was understood by superimposing the permanent water body layer to the extreme crisis day. In Fig. 3f, the black colour appearance depicts the permanent water body, and the red colour appearance depicts the flood condition on 26th July 2019.

The total built area of the district was estimated to be 361 km^2. It is observed and from the Built Area Inundation Map, Fig. 4, that total inundated built area during this period was estimated to be 78 km^2 on an extreme day. In Fig. 4, the blue colour appearance depicts the water layer, and red is the intersection layer between water and built area.

Overall, the inundation is observed to be prolonged in the North-Eastern and South-Eastern parts of the district. On 14th July, the condition appears severe towards North-Eastern parts of the district, whereas the Western parts to Barpeta town seem to remain dry. On 21st July, the condition of North-Eastern parts seems to be improved, but the areas West to Barpeta town seem to be inundated. On the other hand, on 26th July, the spatial extent of inundation was spread to almost all parts of the district. However, the situation reflects to be improved on 7th August 2019.

5 Conclusion

The study tried to establish that satellite imagery like SAR images which can challenge the obstacles of clouds, fogs, and mists, etc. and can be very potential to identify and generate near real-time cost-effective flood mapping and monitoring. Its integration with GIS can produce approximate quantitative values for tentative inundation extents and the impact of the flood on living and non-living objects. Extraction of such crucial information can help in quick and efficient decision making for the quick response and better management and minimize the impact and risk of disastrous situations.

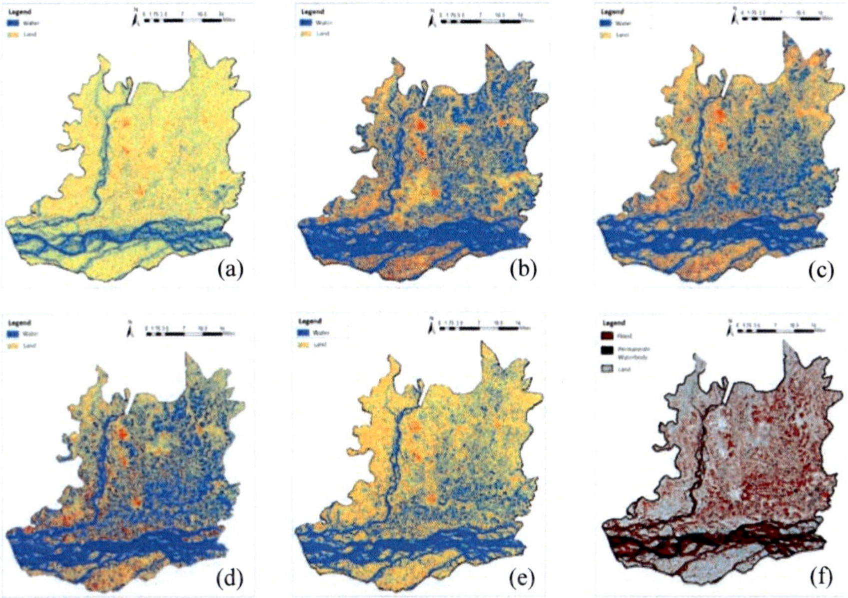

Fig. 3 Different situations of the district. **a** The archived situation with the permanent water bodies; **b** 14th July 2019; **c** 21st July 2019; **d** 26th July 2019; **e** 7th August 2019; **f** the superimposition of archived and extreme flood condition

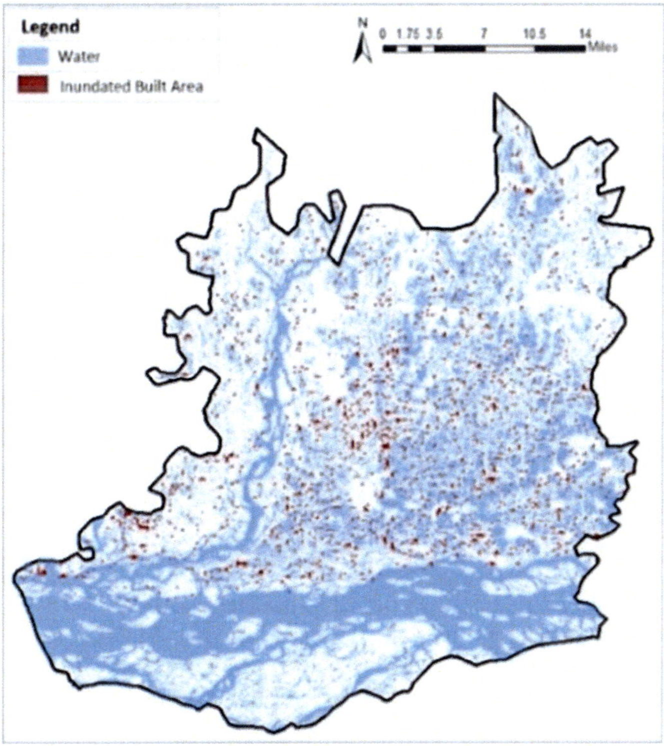

Fig. 4 The built area inundation map

Acknowledgements. We thank ESA Copernicus for the open access of Sentinel-1A SAR data and ESA Sentinel Toolbox Exploitation Platform (STEP) for open access of Sentinel Application Platform (SNAP) and Sentinel-1 Toolbox. We also want to thank Assam State Disaster Management Authority (ASDMA) for daily flood report data for the crisis period.

Conflict of Interest. There is no conflict of interest.

References

1. Bhatt CM, Rao GS, Begum A, Manjusree P, Sharma SVSP, Prasanna L, Bhanumurthy V (2013) Satellite images for extraction of flood disaster footprints and assessing the disaster impact: Brahmaputra floods of June–July 2012, Assam, India. Curr Sci 104(12):1692–1700. JSTOR. https://www.jstor.org/stable/24092606
2. Bhatt CM, Srinivasa Rao G, Manjushree P, Bhanumurthy V (2010) Space based disaster management of 2008 Kosi floods, North Bihar, India. J Indian Soc Remote Sens 38(1):99–108. https://doi.org/10.1007/s12524-010-0015-9
3. Chini M, Pelich R, Pulvirenti L, Pierdicca N, Hostache R, Matgen P (2019) Sentinel-1 InSAR coherence to detect floodwater in urban areas: Houston and Hurricane harvey as a test case. Remote Sens 11(2):107. https://doi.org/10.3390/rs11020107
4. Cian F, Marconcini M, Ceccato P, Giupponi C (2018) Flood depth estimation by means of high-resolution SAR images and lidar data. Nat Hazard 18(11):3063–3084. https://doi.org/10.5194/nhess-18-3063-2018
5. DeVries B, Huang C, Armston J, Huang W, Jones JW, Lang MW (2020) Rapid and robust monitoring of flood events using Sentinel-1 and landsat data on the Google Earth Engine. Remote Sens Environ 240:111664. https://doi.org/10.1016/j.rse.2020.111664
6. Doocy S, Daniels A, Packer C, Dick A, Kirsch TD (2013) The human impact of earthquakes: a historical review of events 1980-2009 and systematic literature review. PLoS Curr 5. https://doi.org/10.1371/currents.dis.67bd14fe457f1db0b5433a8ee20fb833
7. Gaurav K, Sinha R, Panda PK (2011) The Indus flood of 2010 in Pakistan: a perspective analysis using remote sensing data. Nat Hazards 59(3):1815–1826. https://doi.org/10.1007/s11069-011-9869-6
8. Gomathi M, Priya MG, Gowda CC, Krishnaveni D (2019) Flood inundation mapping for using Sentinel-1 SAR data for Assam during 2018. Res Rev J Space Sci Technol 8(2). ISSN: 2321-2837 (Online), ISSN: 2321-6506 (Print)
9. Guha-Sapir D, Hoyois P, Wallemacq P, Below R (2016) Annual disaster statistical review 2016: The numbers and trends by Centre for Research on Epidemiology of Disasters (CRED)
10. Kowalzig J (2008) Climate, poverty, and justice: what the Poznań UN climate conference needs to deliver for a fair and effective global climate regime, 32
11. Li Y, Martinis S, Plank S, Ludwig R (2018) An automatic change detection approach for rapid flood mapping in Sentinel-1 SAR data. Int J Appl Earth Obs Geoinf 73:123–135. https://doi.org/10.1016/j.jag.2018.05.023
12. Rahman MdR, Thakur PK (2018) Detecting, mapping and analysing of flood water propagation using synthetic aperture radar (SAR) satellite data and GIS: a case study from the Kendrapara District of Orissa State of India. Egypt J Remote Sens Space Sci 21:S37–S41. https://doi.org/10.1016/j.ejrs.2017.10.002
13. Sanyal J, Lu XX (2004) Application of remote sensing in flood management with special reference to monsoon Asia: a review. Nat Hazards 33(2):283–301. https://doi.org/10.1023/B:NHAZ.0000037035.65105.95

14. UNISDR, CRED (2017) Economic losses, poverty & disasters 1998–2017. CRED, UNISDR. https://www.preventionweb.net/files/61119_credeconomiclosses.pdf
15. Wood M, de Jong SM, Straatsma MW (2018) Locating flood embankments using SAR time series: a proof of concept. Int J Appl Earth Obs Geoinf 70:72–83. https://doi.org/10.1016/j.jag.2018.04.003
16. Yuan T, Lee H, Jung HC, Aierken A, Beighley E, Alsdorf DE, Tshimanga RM, Kim D (2017) Absolute water storages in the Congo River floodplains from integration of InSAR and satellite radar altimetry. Remote Sens Environ 201:57–72. https://doi.org/10.1016/j.rse.2017.09.003

Chapter 18
Analysis of Large Dam Storage Capacity and Its Effect on Water Demand Management in India

Upadhyay Mudita and M. A. Sherly(✉)

Coca-Cola Department of Regional Water Studies, TERI School of Advanced Studies, Vasant Kunj, New Delhi, India
{mudita.upadhyay,sherly.ma}@terisas.ac.in

Abstract. Dams are considered to be temples of modern India as they serve monumental purposes like water conservation, drought and flood control, irrigation, energy requirements, and food security; however, they also have major socioeconomic and environmental drawbacks. Currently, significant parts of India suffer from agricultural drought, and we need to combat this through critically analyzing our water resource management policies. Our policies are centered on large dams. However, the study shows that large dams are not fulfilling the irrigation requirements of different states in India. We need to change our path that is a large dam-driven and should implement cost-effective, environment-friendly, and socially acceptable measures to conserve water and alleviate water scarcity.

Keywords: Large dams · Irrigation · Effective water storage capacity · Rainfall · Trend analysis · National water policy

1 Introduction

Water is the most vital resource for the survival of all living beings. But it is not uniformly distributed across different regions of the world. Some regions are blessed with water abundance, while on the other hand, some other regions suffer from extreme water scarcity. Regions with water-scarce conditions are susceptible to physical drought, and the areas with high water availability face the issue of water resources management and end up with operational drought. Dams are considered the most crucial factor in alleviating the impulses of water scarcity, drought, and floods. It can store water and help in situations when there is an urgent need to release the water. It can supply water as per the demand of the population. Rivers are a significant and vital source of freshwater supply, so human beings have come up with the idea to build dams to divert and use water from the rivers. Large dam construction is considered an essential factor in alleviating the impulses of water scarcity, drought, and floods. A large dam may be at the height of 15 m or greater from the lowest foundation to crest or a dam between 5 and 15 m, impounding more than 3 million cubic meters (as per the definition by ICOLD—International Commission on Large Dams; [8]).

R. K. Bhattacharjya et al. (eds.), *Sustainable Water Resources Management*,
Advances in Sustainability Science and Technology,
https://doi.org/10.1007/978-981-16-7535-5_18

Worldwide, dams are constructed to harness water for mitigating the effects of drought, leading to a significant contribution to the net irrigation area. India is one of the countries with major water scarcity, with prevalent drought for the past few years. Water availability per capita is decreasing with the rapidly increasing population. Competition for the water resources will lead to severe national food security issues as well as affect the livelihoods of the farmers. In India, annual precipitation is highly variable; hence, it is crucial to develop better water management policies and critical analysis of current and past water resource management projects. Major negligence and deficiency of operation and maintenance of water resources development projects lead to water stress from time to time in numerous regions across the country. The main challenge for India is its escalating water demand combined with socioeconomic development. India's growing population has put an enormous burden on India's water resources. Water demand is progressively increasing across all regions. Assessment of water demand with existing and accessible supply will be critical for the development of India to encounter its growing water demands.

During the time of independence, there were few large dams as compared to at present. Large dams have grown to over 4000, and most of them were built in two decades, i.e., 1970–1990. Large dam construction was one of the major initiatives taken by the Indian Government to tackle water resources management issues. This investment was focused upon the increment in the irrigation and command area. Cullet and Gupta [4] state that large dams are meant to provide food security, fulfill energy, and drinking water requirements. The major part of the country has been facing severe droughts since 1871, which is causing harm to the wider population of India agriculturally, economically, and socially. Large dam water storage augmentation was considered as the instrument to reduce and mitigate the impacts of drought. However, this idea is currently taking a different direction due to the associated socioeconomic and environmental impacts. A study conducted by Meher [6] states that the Hirakud Dam was unable to fulfill the irrigation requirements and proven to be a big failure as a multipurpose dam that led to significant socioeconomic impacts. Baboo [2] states that due to the Hirakud construction, people had to migrate, and they suffered from a stigma called "reservoir outees."

This study analyzes the role of the effective water storage capacity of the large dams of India on the water demand by increasing population, rainfall, and net irrigated area through canals and tanks for the 28 states and three union territories of India for the period of 1970 to 2019. Duflo and Pandey [5] came up with an analysis that showed that dams led to an increase in irrigation area and agricultural productivity. Shah and Kumar [9] concluded that increasing the dam storage capacity will be necessary for economic growth and development. This study analyses the rainfall pattern across India within the context of drought and the effect of the increased effective water storage capacity of the dam on the net irrigated area by tanks and canal, thus, investigates if the increased water storage capacity of large dams has any positive impact on irrigation.

Evolution of Dams in India

The dams have been playing an important role in harnessing the river waters for accelerating socioeconomic growth and mitigating the distress of a large population. They contribute significantly in providing water supply for municipal and industrial purposes.

Moreover, dams have been instrumental for modern India by improving irrigation and generating hydroelectricity apart from acting as flood control structures. The foremost use of dams for hydroelectricity generation was around 1890, and by 1900, numerous large dams had been commissioned for construction around the world [1]. Dam and canal irrigation was prevalent in India long since the British colonial era. Still, after independence in India, it took a boom as Prime Minister Jawaharlal Nehru's vision determined that India should be sufficiently independent in food production. He endorsed the dam as "temples of modern India." Duflo and Pandey [5] state that in India, both state and union governments are responsible for the construction of dams. Planning commission sets the goal for water augmentation, and irrigation departments of the state government proposes the dam project and planning commission finally commissions the project. Subsequently, the five-year plans promoted the construction of large dams for improving irrigation and enhancing food productivity, drought mitigation, and flood control.

First, through the five-year plan (1951–1956) Nagarjuna Sagar Dam was built on the Krishna River in Andhra Pradesh, Bhakra Nangal Dam on the Sutlej River in Punjab, and Hirakud Dam on the Mahanadi River in Odisha. The major portion of the budget in first five-year plan was allocated to irrigation and energy, i.e., 27.2%, which led to the construction of large dams in India. In all five-year plans, significant importance was given to the establishment of additional irrigation potential. Bandyopadhyay [3] states that the five-year plan spent about 20,000 crore rupees (15% of the total plan expenditure) on dam construction, which led to the birth of 1554 large dams at the various stages of implementation. "Till the end of the Eighth five-year plan (March 1997), India has spent Rs. 1,378,088.1 billion at constant 1996–97 prices and Rs. 580,851.3 billion at the current price levels on major and medium (M & M) irrigation projects alone. This includes expenditure on Command Area Development (CAD), which is mostly in the command areas of M&M projects" [10]. In 1965, the Green Revolution was introduced in India with the objective to expand the sources of irrigation that led to the construction of many large dams from 1968 to 1979. In 1974–75, the Government of India introduced the command area development program to narrow the gap between irrigation potential created and utilized. This led to the laying of pipes, construction of new dams, canals, and tanks to provide water to the agricultural fields.

Role of National Water Policy

In 1987, National Water Policy (NWP) was introduced after the major 1987 drought. The main objectives of NWP are the promotion of conjunctive use of water from surface and subsurface sources, supplemental irrigation, and water-conserving crop pattern and irrigation and production technologies (GOI 1987). NWP has called for raising the canal water rates and promoting user participation in canal management. Further, NWP also aimed at improvement in irrigation facility and command area development. It also focused on food security, increase in crop productivity, mitigation of drought, and control of flood. It further led to the construction of large dams as they are considered to increase irrigation, which can lead to an increase in crop productivity. After the implementation of NWP 1987, new challenges arose in the water sector, which demanded a review of NWP as the policy was criticized because of its encouragement to the construction of big dams. Pandit and Biswas [7] state that the policy was influenced by the "big dam lobby" that rendered it ineffective. Accordingly, the revised NWP-2002 was adopted by the National Water Resources Council on April 1, 2002. There was a better realization for a paradigm shift from water resources development to water resources management by restructuring and strengthening existing institutions.

Increasing population, intensifying needs of a rapidly developing nation as well as the implications of the climate change impact prompted a further review of the National Water Policy 2002. The adopted NWP 2012 was released during India water week 2013. NWP 2012 highlighted on treating water, over and above the pre-emptive need for safe drinking water and sanitation, as an economic good. The Government of India has introduced the National Water Mission as one of the eight National Missions under the National Action Plan on Climate Change. The Union Cabinet approved (on April 6, 2011) the comprehensive Mission Document for National Water Mission (NWM). The main objective of NWM is preservation of water, diminishing wastage and safeguarding its more justifiable circulation both across and within the states through integrated water resources development and management, which promotes the idea of robust water resource development while encouraging the exhaustive assessment of dam construction projects.

2 Materials and Methods

Initially, the data for different attributes have been collected from various sources (Table1). The drought years during 1970–2019 are shown in Table 2 along with the relevant data on annual rainfall, average large dam storage capacity. Average irrigated area from tanks and canals, and total population. Descriptive statistics have been performed on the effective storage capacity of large dams for the period of 50 years. Decadal analysis has been performed for the states, and correlation coefficients have been calculated for effective storage capacity of large dams versus population, population versus year, rainfall versus year, rainfall versus net irrigated area by canals and tanks, net irrigated area by canals, and tanks versus year. The calculated correlation coefficients have been demonstrated with the help of maps as shown in Fig. 1.

Table 1 Source of the various data used in the study

Variables	Data sources
Population data	Census of India
Large dam effective storage capacity	Central Water Commission, National Register of large dams 2019
Net irrigated area from canals and tanks (all social groups)	Directorate of Economics and Statistics, Dept. of Agriculture, Cooperation and Farmers Welfare, Ministry of Agriculture, Govt. of India
Actual rainfall (mm)	India Meteorological Department

Table 2 Effect of increment in dam storage capacity on drought years in India from 1970 to 2019

Major meteorological drought year from (1970 to 2019)	Annual rainfall (mm)	Average large dam storage capacity in (thousand cubic meters)	Average irrigated area from tanks and canals in (ha)	Total population in million
1972	910.41	2,757,642.41	875,581.53	548.16
1974	1031.55	3,187,719.09	875,581.53	(Census 1971)
1979	999.60	3,857,075.61	592,040.73	
1982	1057.74	4,026,195.52	613,588.66	683.33
1985	1101.09	4,484,322.52	567,181.50	(Census 1981)
1986	1089.44	4,551,320.58	567,181.50	
1987	1052.50	4,729,237.86	567,181.50	
2002	895.54	5,666,841.62	559,000.00	1028.74
2009	929.96	6,008,286.46	632,633.33	(Census 1981)
2014	1074.81	6,120,728.79	677,466.66	1210.19
2015	1089.66	6,121,826.65	677,466.66	(Census 2001)

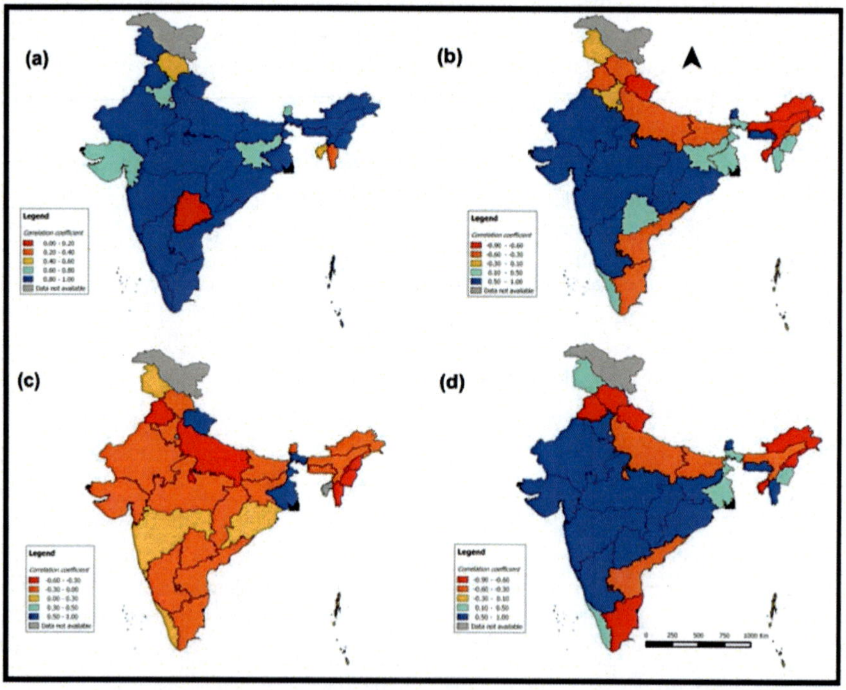

Fig. 1 Correlation analysis of the various factors **a** Effective water storage capacity and population. **b** Effective water storage capacity and net irrigated area from tanks and canals. **c** Rainfall trend. **d** Trend of net irrigated area from tanks and canals

3 Results and Discussion

It has been observed that the top ten states with the highest effective water storage capacity of large dams are Maharashtra, Karnataka, Odisha, Telangana, Himachal Pradesh, Uttar Pradesh, Madhya Pradesh, Gujarat, Kerala, and Rajasthan. This analysis has been performed through descriptive statistics on the effective water storage (billion cubic meters). It has also been found that the highest variation in effective water storage augmentation occurred from 1970 to 1990 through coefficient of variation. Most of the states and union territories showed positive correlation between effective water storage of large dams with the increase in population except for Telangana, Himachal Pradesh, Mizoram, and Tripura (Fig. 1a). In comparison of the net irrigated area by tanks and canals to the effective water storage capacity of large dams, it has been observed that north, southeast, and northeast regions of the country are negatively correlated, while the west and central part of India has positive correlation (Fig. 1b). A decreasing trend in annual rainfall has been observed in general except for Jammu and Kashmir, Uttarakhand, Kerala, Maharashtra, Andaman and Nicobar, Odisha, and Goa (Fig. 1c). The trend of net irrigated area from canals and tanks (Fig. 1d) shows a similar trend as that of Fig. 1a with Punjab, the rice bowl of India showing a negative trend. This may indicate the increased dependence on groundwater in the State.

4 Conclusion

National Water Policy 1987 implemented by the National Water Resource Council focused upon command area development, which led to the large dam construction in various parts of India. As evident from the results, major large dam construction happened in the 1970s–1990s. The prime purpose of the dam construction was to increase the net irrigation area, and results show a decreasing trend of irrigation through canals and tanks. Canals and tanks are the primary sources of utilizing dam-stored water for irrigation, and this indicates that dam water storage augmentation has not contributed to increasing the net irrigation area in recent times. From our analysis, it is evident that the rainfall trend is decreasing, which reflects upon the barren situation of the large dams in India. Interestingly, Himachal Pradesh is one of the states with the highest effective water storage capacity of large dams. Still, from the results, it is indicated that the trend of area irrigated by tanks and canals is decreasing, which reflects that building a large dam in the high earthquake-prone area is not contributing towards the cause and proves to be high-risk development. The large dam constructed without sufficient planning can be proved to be unsuccessful in delivering benefits. However, dam construction is necessary to satiate the need for growing water demand; hence, it is important that such water resource development projects are commissioned through effective measures to abate negative impacts. Large dams promote the efficiency of irrigation and provide additional benefits like increasing water supplies to remote areas, providing drinking water, help in the reduction of flooding, and have proven to be an efficient means for generation of hydroelectricity. However, on the other hand, building dams without adequate planning for command area development may lead to conflicting interests. Command area development should essentially integrate traditional water harvesting systems already existing in the command area. Water-saving techniques and technologies should be promoted, creating a sense of not only increasing supply but also the regulation of demand. A range of socioeconomic and ecological interventions can be devised to ameliorate the negative impacts of dams, like increasing the participation of various stakeholders before the construction of a dam, assessing the impacts of dam adequately, studying the benefits of ecosystem services and biodiversity, and integrating them into cost–benefit analysis during the planning phase. Lack of robust framework and numerous loopholes in guidelines pave an easy way for blindfolded construction and commissioning of dams. It is essential to learn lessons from past failures like commissioning of dam without appropriate assessment of the geographical area, cost–benefit analysis, and the impacts of dam on socioeconomics and ecology of the area. A strong framework along with strict environment impact assessment (EIA) and public participation with appropriate timelines should be introduced to make future water resource development projects ingenious.

References

1. Altinbilek D (2002) The role of dams in development. Water Sci Technol 45(8):169–180
2. Baboo B (2009) Politics of water: the case of the Hirakud dam in Orissa, India. Int J Sociol Anthropol 1(8):139–144
3. Bandyopadhyay J (1987) Political ecology of drought and water scarcity: need for an ecological water resources policy. Econ Polit Wkly 2159–2169
4. Cullet P, Gupta J (2009) India: evolution of water law and policy. In: The evolution of the law and politics of water. Springer, Dordrecht, pp 157–173
5. Duflo E, Pande R (2007) Dams. Q J Econ 122(2):601–646
6. Meher R (2011) Big dam, big failures: a study of the canal irrigation system and the deprived tail-end farmers in the Hirakud command area of Orissa, India. J Asian Afr Stud 46(4):422–438
7. Pandit C, Biswas AK (2019) India's national water policy: 'feel good document, nothing more. Int J Water Resour Dev 35(6):1015–1028
8. Proske D (2018) Comparison of dam failure frequencies and failure probabilities. Beton-und Stahlbetonbau 113:2–6
9. Shah Z, Kumar MD (2008) In the midst of the large dam controversy: Objectives, criteria for assessing large water storages in the developing world. Water Resour Manage 22(12):1799–1824
10. Thakkar H (2000) Assessment of irrigation in India. World Commission of Dams

Chapter 19
Impact of Nodal Points on River Morphology of Brahmaputra River

Dipsikha Devi[1(✉)], Dipima Sarma[2], Dhruba Jyoti Sarmah[1], Arup Kumar Sarma[1], and Rajib Kumar Bhattacharjya[1]

[1] Department of Civil Engineering, Indian Institute of Technology Guwahati, Guwahati, Assam, India
{aks,rkbc}@iitg.ac.in

[2] Climate Change Cell, Forest Department, Government of Assam, Guwahati, Assam, India

Abstract. The Brahmaputra River is one of the major trans-boundary rivers passing through China, India, and Bangladesh. It is one of the youngest rivers of the world characterized by braiding and susceptible to continuous channel migration and avulsion. This study elucidates the impact of the nodal points on river morphology. In this study, toposheets of 1973–94 and four sets of satellite imageries, namely Landsat MSS, LISS-I, LISS-III, and LISS-IV, pertaining to the years 1976–80, 1993–95, 2003–04, 2008–11, and 2016–17, respectively, were used. The banklines were digitized for each dataset, and the widths of the channel section at an interval of 10 km were estimated from Sadiya to downstream of Dhubri. The nodal points were then identified by superimposing banklines of different decades and by observing the locations that have not undergone changes. For the convenience of the observation,the widths of each of the river sections were plotted symmetrically about the centerline. The major nodal points identified were the most constricted width of the Brahmaputra River, where the bridges are located, which includes Bogibeel bridge in between chainage 70–80 km, Koliabhomora bridge in between chainage 280–290 km, Saraighat bridges in between 420–430 km, and Naranarayan Setu in between chainage 530–540 km. Another nodal point was observed at chainage 150 km at Majuli, which is the largest inhabited river island in the world. This nodal point can be regarded as an artificial one created by bank protection measures taken on both banks of the river. A decadal comparative analysis of erosion and deposition at the upstream and downstream of 10 km of these nodal points was conducted from all the dataset. From the analysis, an interesting fact observed is that the erosion at downstream of all the nodal points is continuing, and the decadal rate of erosion has come down significantly with time. A similar trend has been observed in the deposition process also.

Keywords: Brahmaputra · Nodal points · River morphology · Satellite imageries · Bank protection measures · Erosion

R. K. Bhattacharjya et al. (eds.), *Sustainable Water Resources Management*, Advances in Sustainability Science and Technology,
https://doi.org/10.1007/978-981-16-7535-5_19

1 Introduction

River morphodynamics is a consequence of channel dimension, gradient, and a process of erosion and deposition [1]. The alluvial rivers like the Brahmaputra and the Ganges show various seasonal erratic behaviors [2]. Thus, the planform of these rivers is extremely dynamic, and several temporal variations can be observed. Human activities like damming, channelization, urban effects, industrialization, etc., are some of the factors that can be attributed to the change of the river morphology to a great extent [3, 4]. Recent advancement in space science has made possible the improvement in remote sensing and GIS technology which is currently in vogue as an efficient application in the field of water resources. Remote sensing and GIS technology have helped to extract information where field survey is inaccessible. Aerial photographs and satellite imageries are powerful means that can detect the morphological changes of a river. The remote sensing and field survey data can also be used for calibration and validation of various planform results generated using numerical models.

Nodal points are important features which have a great impact on river morphology. Nodal points of a river are those localized points where the river converges upstream and diverges downstream. Several pieces of research were conducted since the late 90s, and several nodal point relationships have been studied to understand the distribution of sediments at channel bifurcation [5]. Wang et al. [6] proposed a nodal point relationship in one-dimensional network morphodynamic model which states that

$$\frac{S_1}{S_2} = \left(\frac{Q_1}{Q_2}\right)^K \left(\frac{B_1}{B_2}\right)^{1-K} \tag{1}$$

where $S1$, $S2$, $Q1$, and $Q2$ are the sediment and water discharge at the downstream channels, and $B1$ and $B2$ are the two widths of the two channels.

Study Area

The Brahmaputra basin extends between the geographical bounds of 26.03°N to 27.48°N latitude and 90.00°E to 95.00°E longitude. It is a trans-boundary river that flows through China, India, and Bangladesh. It is the youngest of all the major rivers in the world [7]. During its flow through the hilly region of Arunachal Pradesh, the river is concise within the hills, but as soon as it enters the alluvial plains of Assam, fanning out of the river can be observed due to the sudden change in the slope [8]. Alluvial rivers are mostly associated with flood, erosion along with sedimentation. The Brahmaputra River is not an exception in these problems. Every year flood and erosion problem is a major cause of concern in Assam, Arunachal Pradesh, and other states of North East India. In this study, the Brahmaputra River flowing through Assam has been considered, which includes Sadiya, the easternmost part of Assam to Dhubri, the westernmost part in Assam (Fig. 1).

2 Materials and Methods

The nodal points were identified as the most constricted width of river Brahmaputra extracted using GIS which has not undergone any changes for around 50 years. For the

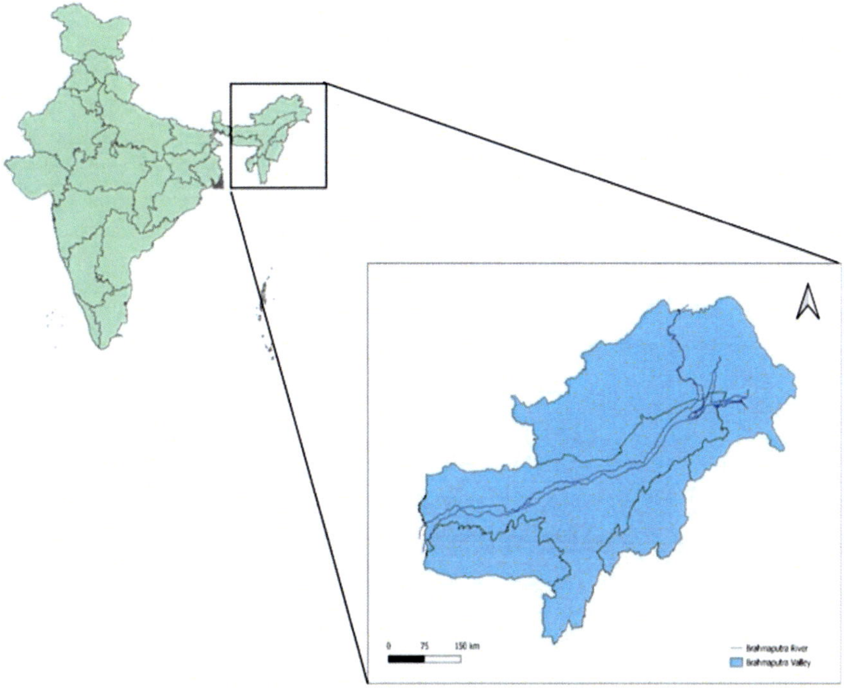

Fig. 1 Study area

analysis, the geospatial data were used which include one set of SOI toposheets of 1973–74 and five sets of satellite data for the period Landsat MSS 1976-80 (80 m resolution), IRS 1B LISS1 1993-95 (72.5 m resolution), IRS P6 LISS3 2003-04 (23.5 m resolution), Resourcesat-1 LISS 3 2008-11 (23.5 m resolution), and Resourcesat-2 LISS 4 2016-17 (5.8 m resolution) at around 10 years interval. The toposheets were then georeferenced, subset, and mosaicked using Erdas Imagine 2015. For the satellite imageries, layer stacking and mosaicking were done in Erdas Imagine 2015.

Bankline Delineation

Identifying and delineating river banklines is an important assessment to understand the channel morphology, which is often dependent on the resolution and skill of the interpreter. A comparative analysis between automatic delineation and manual delineation inferred that the manual digitization method is easy for the interpreter to demarcate the river banklines due to visual differences between the lateral sandbars and the bank landmass [9]. The entire Brahmaputra River system, with the confluence of important tributaries from Sadiya to Dhubri (near Bangladesh border), was delineated from satellite imagery. For river bankline delineation, polyline shapefiles were created for each time period, and digitization was done following standard digitizing procedure manually. The delineated banklines were then analyzed to quantify the channel dimension along with the erosion process along the banks.

Channel Dimension

For the analysis of the channel dimension of the river Brahmaputra, the entire river from Sadiya to Dhubri had been divided into sections of 10 km reach, and the width of the river at each section for all the study periods was measured. The lateral spreading of the river from its center line was measured for each of these sections, and a plot of the width against each section was plotted as shown in Fig. 2. Five major nodal points were identified from this plot.

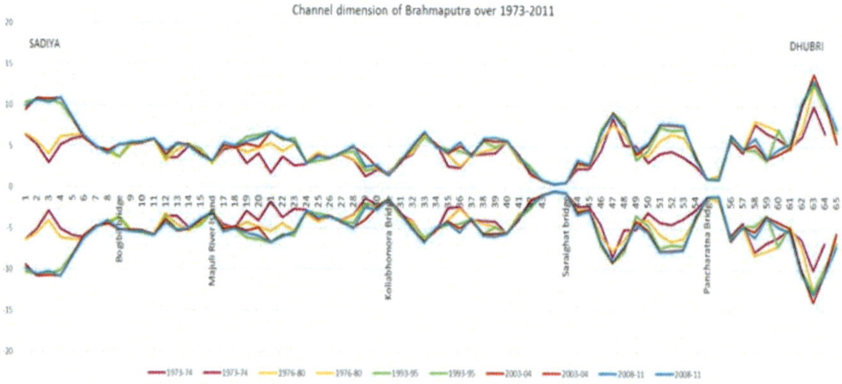

Fig. 2 Channel dimension of Brahmaputra River over 1973–2011

Erosion–Deposition Process

A comparative study has been carried out using the satellite datasets of 1973–74, 1976–80, 1993–95, 2003–04, 2008–11, and 2016–17. The overall erosion–deposition phenomenon considering the left bank and right bank was analyzed for 10 km upstream of the nodal points and 10 km downstream of the nodal points. The right and the left bank of the river Brahmaputra were digitized manually in ArcGIS. The delineated temporal banklines were superimposed following the generation of the polygon shapefiles to estimate the erosion–deposition process of 10 km upstream and downstream of the nodal points. The areas of the polygons were then calculated to quantify the amount of erosion and deposition along the banks. Figures 3 and 4 show the erosion and deposition, respectively, at 10 km upstream and downstream of five different major nodal points.

3 Results and Discussion

From Fig. 2, the major nodal points identified were the most constricted width of the Brahmaputra River. Except for one, these are the points where the bridges are located. The bridges include Bogibeel bridge in between chainage 70–80 km (section 9), Koliabhomora bridge in between chainage 280–290 km (in between sections 30 and 31), Saraighat bridges in between 420–430 km (section 44), and Naranarayan Setu in between

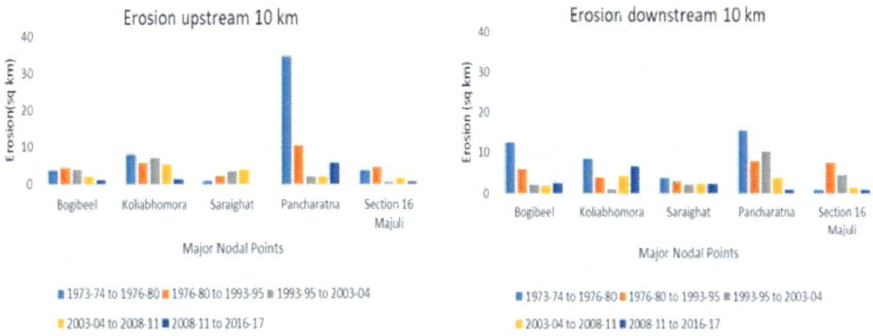

Fig. 3 Erosion at 10 km upstream and downstream of the identified major nodal points

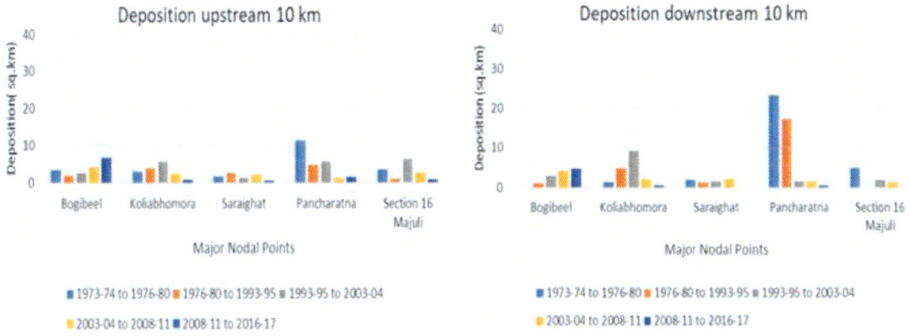

Fig. 4 Deposition at 10 km upstream and downstream of the identified major nodal points

chainage 530–540 km (section 55). Another nodal point was observed at chainage 150 km at Majuli, which is the largest inhabited river island in the world. From the erosion–deposition analysis, it can be seen that although the erosion is continuing after the construction of the bridges, the rate of decadal erosion is decreasing with time in most of the cases. The erosion at upstream can be attributed to the sudden rise and fall of the water level due to afflux and recession during flood time. While erosion at downstream can be associated to the high energy of emerging flow due to the heading up of water at upstream and thus resulting fanning out at downstream. The reduction in stream power due to a progressive decrease in bed slope resulting of fanning out of the river at downstream can be one reason for the decrease in bank erosion rate downstream. Moreover, the dominant influence of various river training works like guide bunds and spurs that were implemented on both the banks near the nodal points cannot be ignored. The nodal point corresponding to section 16 can be regarded as an artificial one created by bank protection measures taken on both banks of the river.

4 Conclusion

The satellite-based study carried out using data of the last five decades has shown that erosion at downstream of nodal point has a decreasing trend in most of the cases. However, the erosion at upstream, though it has a decreasing trend in most of the cases, has not shown a definite pattern. Similarly, the trend of deposition rate is also not very clear. This is because of the fact that the erosion and deposition phenomenon is also governed by various other factors, which include the formation of mid-sand bars, the effect of river training works, and variation in flow and sediment load scenario. The analysis of nearly 50 years of data has shown a reducing trend in respect of bank erosion at downstream of these nodal points. But for a highly braiding and complex river like Brahmaputra, it is difficult to jump into a conclusion that the river is reaching a stable state. This is due to the fact that the change in the sediment influx and discharge scenario may disturb the present configuration of these studied reaches, which are, in some cases, appearing to approach a stable condition. A more detailed study is necessary to evaluate the effects of river training work on the erosion and deposition pattern of the river.

Acknowledgements. We acknowledge Central Water Commission (CWC), Ministry of Jal Shakti, Department of Water Resources, Government of India, for providing funding and support of the project named 'Morphological Studies of Rivers Brahmaputra, Subansiri and Pagladia using Remote Sensing Technique'.

Conflict of Interest. None.

References

1. Church M, Ferguson RI (2015) Morphodynamics: rivers beyond steady state. Water Resour Res 51:1883–1897
2. Goswami DC (1985) Brahmaputra River, Assam, India: physiography, basin denudation and channel aggradation. Water Resour Res 21:959–978.41
3. Horton RE (1932) Drainage basin characteristics. Trans Am Geophys Union 13:350–361
4. Leopold LB, Maddock T (1953) Hydraulic geometry of stream channels and some physiographic implications. U.S. Geol Surv Prof Pap 272:57
5. Islam GMT, Kabir MR, Nishat A (2006) Nodal point relation for the distribution of sediments at channel bifurcation. J Hydraul Eng 132(10):1105–1109
6. Wang ZB, Fokkink RJ, de Vries M, Langerak A (1995) Stability of river bifurcations in 1D morphodynamic models. J Hydraul Res 33(6):739–750
7. Sarma JN (2004) An overview of the Brahmaputra River system, The Brahmaputra basin water resources. In: Singh VP, Sharma N, Shekhar, Ojha P (eds) Water science and technology library, pp 17–23
8. Report on Morphological Study of Brahmaputra River using Remote Sensing Technique (2018) CWC, Ministry of Jal Shakti, Department of Water Resources, Government of India
9. Sarma AK, Sarma D 2018) A comparative study of remote sensing based bank line delineation techniques for an erosion-prone Braided River. In: Conference on next frontiers in civil engineering sustainable and resilient infrastructure, IIT Bombay

Chapter 20
Stability Analysis of Riverbank Erosion

Snigdha Kalita(✉) and P. K. Khaund

Civil Engineering Department, Jorhat Engineering College, Jorhat 785007, India

Abstract. Riverbank erosion is a hydraulic process that is caused due to riverbank instability. The erosion process is often observed in alluvial rivers. The mighty river Brahmaputra in Assam, having alluvial riverbanks, is mostly unstable. A comparative study of the stability analysis of different riverbanks in upper Assam including Majuli Island is represented herein. Soil parameters such as grain size analysis, plasticity index, shear parameters, and permeability are studied for the analysis of bank stability and to find the critical angles for each case. Different soil samples are collected from the Dikhow, Dhansiri, and Lohit river banks and are tested in the laboratory to find out the geotechnical and engineering properties. Culman type of stability analysis for steep, homogeneous, and cohesive river banks proposed by Osman and Thorne in 1988 is selected for the study. The analysis is done using the standard formulas that are based on the results obtained from different laboratory tests. Critical bank angles are calculated from the analysis along with the factor of safety. The work can act as a vital keyway for taking the precaution against the failure of the riverbanks and thereby to mitigate flooding in the rainy season.

Keywords: Riverbank erosion · Culman-type stability analysis · Critical bank angles · Factor of safety

1 Introduction

The riverbank erosion is caused in the matured stage of a river. Continuous water oscillations in the river weaken the riverbanks leading to massive erosion. The height of the riverbank, bank geometry, and bank material property is the key factor that affects the stability of a riverbank. The river Brahmaputra originates in the Northern Himalayan area and flows through China, India, and Bangladesh and finally reaches the Bay of Bengal, having more than twenty tributaries in its North and South banks. In Assam, the Brahmaputra and its tributaries set a classic example of a braided river and are more prone to erosion of the bank as well as the bed. Majuli Island is one of the most prone areas to erosion due to high sediment deposition of the flowing water of the river. Continuous river bank erosion creates a severe threatening to the people of Assam. Therefore, to mitigate such severe threats, the study on stability analysis of the river Brahmaputra and its tributaries in Assam is essential.

R. K. Bhattacharjya et al. (eds.), *Sustainable Water Resources Management*,
Advances in Sustainability Science and Technology,
https://doi.org/10.1007/978-981-16-7535-5_20

Based on the Culman type of stability analysis concept, a comparative study has made to analyze the stability of Brahmaputra riverbank in different locations along with its three southern tributaries in the upper Assam reach based on the critical angles.

2 Materials and Methods

The material used for the study is the bank material, i.e., soil samples of the riverbanks. Six soil samples were collected by core cutter method (According to IS 2720 (Part 29)) from the banks of the river Dikhow, Dhansiri and Lohit. Three samples were collected from Majuli Island in Brahmaputra River in Jorhat district, India.

Geotechnical properties and engineering properties of the samples were tested in the laboratory of Civil Engineering Department, Jorhat Engineering College. Various tests such as Atterberg's limit test, maximum dry density, optimum moisture content, shear parameters (cohesion and angle of internal friction), and coefficient of permeability using falling head test are performed as per different parts of IS 2720.

Standard formulas given by Osman and Thorne in Culman-type of stability analysis method [2] for homogeneous and isotropic riverbanks were used to find out the factor of safety. The factor of safety is given as

$$\text{FOS} = \frac{\text{Resisting force}}{\text{Driving force}} \tag{1}$$

$$\text{The resisting force is, } F_R \frac{(H-y)C'}{\sin\beta} + W_t \cos\beta \, \tan\varphi' \tag{2}$$

$$\text{Driving force is given by, } F_D = W_t \sin\beta \tag{3}$$

$$\text{Weight of the failure block, } W_t = \frac{\gamma}{2}\left(\frac{H^2 - y^2}{\tan\beta} - \frac{H'^2}{\tan i}\right) \tag{4}$$

$$\text{Depth of cracking, } y = KH \tag{5}$$

$$\text{Failure plan angle, } \beta = \frac{1}{2} \times \left[\tan^{-1}\left\{\left(\frac{H}{H'}\right)^2 \left(1 - K^2\right) \tan i\right\} + \varphi\right] \tag{6}$$

Here, K is dependent on the angle of internal friction, γ is the submerged unit weight, H is the initial bank height, c' is effective cohesion, and ϕ' is the effective angle of internal friction. Total height is calculated by adding initial bank height H and degradation depth Δz. H' is taken as half of the total height of the riverbank. The field data of the riverbank such as degradation depth and bank height are collected from the Water Resource Department, Govt. of Assam for the analysis.

3 Results and Discussion

In the present study, the soil samples are grouped as silty sand (SM), clayey sand (SC), silt with low compressibility (ML) and clay with low compressibility (CL) type.

From the stability analysis in the current study using Culman-type analysis method, it is found that the river Dhansiri is stable up to critical angle 78.60° and Dikhow river is stable up to 76.25°. Lohit River is found to be most stable having a critical angle of 80° with ML type of soil. The Brahmaputra in Kamalabari with CL type of soil is stable up to 78.70°, and in Salmara, it is found to be 73°. The Brahmaputra riverbank at Afalamukh with SC type of soil is found to be the most unstable at critical angle 71°.

From the comparison of the present results with the results obtained for the three riverbanks of Khagana Buragohain and P. K. Khaund study [1], it is observed that the Kakodonga riverbank with cllay with intermediate plasticity (CL)-type soil is the most stable one, where the Brahmaputra (Nimatighat) with SM type of soil is stable with a critical angle of 72°. Similarly, Burhidihing riverbank is stable for critical angle 80° with ML soil.

The study informs that the FOS decreases with the increase in failure plane angle for steeper banks. This shows that erosion of riverbank increases for steep riverbanks the most. The FOS increases with the increase in shear parameters.

4 Figures and Tables

In Fig. 1, the considered riverbanks in the method of the Culman-type stability analysis are shown. Before erosion, the bank has initial height H_0 with bank angle *i*. The bank height increases with the bed degradation after erosion as shown in Fig. 1b. The failure plane is passing through the toe making the angle β.

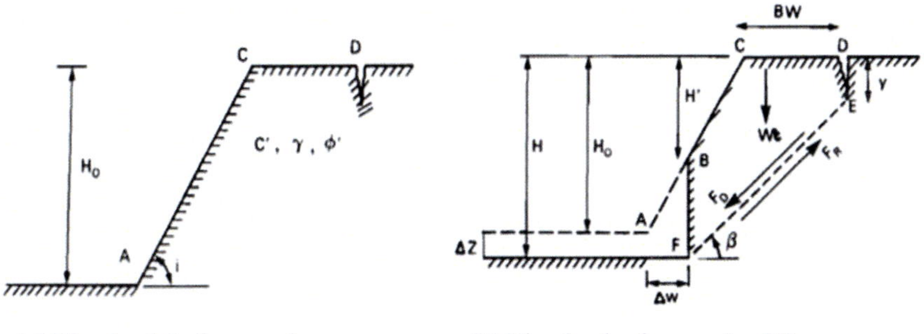

(a) Riverbank before erosion (b) Riverbank after erosion [5]

Fig. 1 **a** Riverbank before erosion. **b** Riverbank after erosion [2]

In Fig. 2, the stability graphs for the considered riverbanks are shown. The factor of safety range from 0.90 to 1.15 except the Kakodonga riverbank which is shown in the other figure. For Kakodonga riverbank, the factor of safety range from 3.1 to 3.9. It is observed that the factor of safety decrease with increase in the bank angle.

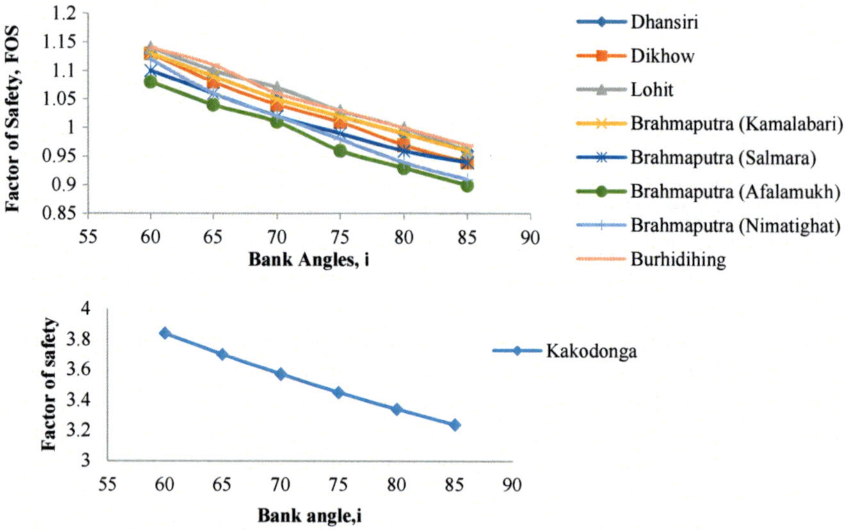

Fig. 2 Stability graphs of the riverbanks

Atterberg's limit test and the plasticity chart, the soil samples are grouped as silty sand (SM), clayey sand (SC), silt with low compressibility (ML), clay with low compressibility (CL) and clay with intermediate compressibility (Tables 1 and 2).

Table 1 Liquid limit, plastic limit, and plastic index of the soil samples

River Bank	LL (%)	Pl (%)	PI
Dhansiri	34.5	28.0	6.5
Dikhow	35.0	24.4	10.6
Lohit	35.0	28.3	6.7
Brahmaputra (Kamalabari)	34.0	23.0	11.0
Brahmaputra (Salmara)	31.0	26.0	5.0
Brahmaputra (Afalamukh)	32.3	24.8	7.5
*Brahmaputra (Nimatighat)	31.0	26.2	4.8
*Burhidihing	32.0	24.0	8.0
*Kakodonga	43.0	20.5	22.5

*The data are taken from the research paper published by Buragohain and Khaund [1] for the comparison with the present study

Table 3 represents the stability parameters such as degradation depth, cohesion, and angle of internal friction, which are used for the stability analysis of the banks. The results for the analysis of the factor of safety are summarized in Table 4.

Table 2 Engineering properties of the soil sample

River bank	Water content (%)	MDD (g/cc)	OMC (%)	Bulk unit weight (g/cc)	Permeability K (cm/s)
Dhansiri	34.5	1.70	16.0	1.75	4.36×10^{-5}
Dikhow	39.0	1.67	19.2	1.83	3.27×10^{-7}
Lohit	35.0	1.68	16.7	1.77	8.85×10^{-5}
Brahmaputra (Kamalabari)	38.0	1.65	18.7	1.83	4.46×10^{-7}
Brahmaputra (Salmara)	32.0	1.62	15.9	1.71	9.26×10^{-5}
Brahmaputra (Afalamukh)	36.0	1.63	13.0	1.77	4.08×10^{-6}
Brahmaputra (Nimatighat)	33.54	1.62	12.0	1.73	1.23×10^{-4}
Burhidihing	34.0	1.72	16.5	1.84	6.93×10^{-5}
Kakodonga	45.54	1.575	23.0	1.83	2.56×10^{-8}

Table 3 Stability parameters of the soil samples

River bank	Degradation depth, ΔZ, (m)	Cohesion, C (kN/m^2)	Angle of internal friction, φ (°)
Dhansiri	0.25	9.0	23.5
Dikhow	0.25	12.0	16.8
Lohit	0.15	9.0	24.0
Brahmaputra (Kamalabari)	0.30	16.5	21.6
Brahmaputra (Salmara)	0.30	6.8	32.2
Brahmaputra (Afalamukh)	0.30	8.0	26.4
Brahmaputra (Nimatighat)	0.30	7.5	32.0
Burhidihing	0.25	10.0	24.23
Kakodonga	0.15	48	20.0

Table 4 FOS calculations for different bank angles

Riverbank	Bank angle, i	Bank height, H_0 (m)	Critical bank angle, i_{cr}	Failure plane angle, β	FOS
Dhansiri (ML)	60°	2.5	78.60°	51.73°	1.13
	65°			52.68°	1.09
	70°			53.57°	1.05
	75°			54.40°	1.02
	80°			55.2°	0.99
	85°			55.98°	0.96
Dikhow (CL)	60°	2.5	76.25°	47.51°	1.13
	65°			48.62°	1.08
	70°			49.65°	1.04
	75°			50.63°	1.01
	80°			51.57°	0.97
	85°			52.48°	0.94
Lohit (ML)	60°	2.5	80.00°	52.02°	1.14
	65°			52.96°	1.10
	70°			53.84°	1.06
	75°			54.67°	1.03
	80°			55.47°	1.00
	85°			56.24°	0.97
Brahmaputra (Kamalabari) (CL)	60°	4	78.70°	50.61°	1.13
	65°			51.60°	1.09
	70°			52.51°	1.05
	75°			53.38°	1.02
	80°			54.21°	0.99
	85°			55.01°	0.96
Brahmaputra (Salmara) (SM)	60°	4	73.00°	56.57°	1.10
	65°			57.43°	1.06
	70°			58.23°	1.02
	75°			58.98°	0.99

(*continued*)

Table 4 (*continued*)

Riverbank	Bank angle, i	Bank height, H_0 (m)	Critical bank angle, i_{cr}	Failure plane angle, β	FOS
	80°			59.70°	0.96
	85°			60.40°	0.94
Brahmaputra (Afalamukh) (SC)	60°	4	71.00°	53.39°	1.08
	65°			54.30°	1.04
	70°			55.15°	1.01
	75°			55.95°	0.96
	80°			56.72°	0.93
	85°			57.46°	0.90
Brahmaputra (Nimatighat) (SM)	60°	4	72.00°	54.99°	1.12
	65°			55.85°	1.06
	70°			56.64°	1.02
	75°			57.39°	0.98
	80°			58.11°	0.94
	85°			58.81°	0.91
Burhidihing (ML)	60°	2.5	80.00°	52.14°	1.14
	65°			53.09°	1.11
	70°			53.96°	1.06
	75°			54.79°	1.03
	80°			55.58°	1.00
	85°			56.35°	0.97
Kakodonga (CI)	60°	2.5	–	49.62°	3.84
	65°			50.64°	3.70
	70°			51.58°	3.57
	75°			52.48°	3.45
	80°			53.34°	3.34
	85°			54.18°	3.24

5 Conclusions

From the study, it is seen that Culman-type method of stability analysis can be applied to the alluvial riverbanks. The analysis is done for steep riverbanks, i.e., bank angle starting from 60° to 85°. The banks are found to have critical angles between 70° and 80°. The method is effective for taking the preventive measures against the failure of the banks in the selected site of the river. But since the work is done at only one site of each river, conclusion cannot be made for the whole river. So, for more accurate results, at least ten sites of each river should be selected for the stability analysis of the riverbanks.

Acknowledgements. I express my sincere gratitude to my guide Dr. P. K. Khaund, Prof. & HOD, Department of Civil Engineering, Jorhat Engineering College, Jorhat, for his continuous support, motivation, and valuable suggestions throughout the preparation of this work.

References

1. Buragohain K, Khaund PK (2019) Study on stability analysis of south bank of River Brahmaputra and its tributaries in the reaches of upper Assam. In: TH-07-41, Preceeding on Indian geotechnical society, Surat chapter
2. Osman AM, Thorne CR (1988) Riverbank stability analysis. I: theory. J Hydraul Eng 114(2)

Chapter 21
Morphological Changes of River Dikrong with Due Emphasis on Effects of Ranganadi Hydroelectric Power Plant

Dhruba Jyoti Sarmah[1], Sanjib Gohain[2(✉)], and Rajib Kumar Bhattacharjya[1]

[1] Department of Civil Engineering, Indian Institute of Technology Guwahati, North Guwahati 781039, India
rkbc@iitg.ac.in

[2] Department of Civil Engineering, Assam Engineering College, Guwahati 781013, India

Abstract. With every new dawn, there have been changes in the spheres of the earth. These changes have many times touched the morphology of a river. Man-made interventions like the construction of a dam also add to changes in the geomorphological characteristics of a river. The shifting of the course of a river due to various reasons is seen to be a common phenomenon that affects the life of local habitats. The advances in aerial photographs, satellite imageries, remote sensing, and GIS technologies significantly help in gathering information from remote areas where field survey is inaccessible, thereby standing to be a fundamental tool for morphological studies of rivers. In the present study, the morphological changes of river Dikrong with due emphasis on the effects of the Ranganadi hydroelectric power plant are tried to find out using multi-temporal satellite data coupled with geospatial techniques. For it, the Landsat datasets were collected, which were digitalized, processed, and analysed using ArcGIS to find the amount of erosion, deposition, changes in channel dimension, and bankline shifting for a reasonable time period. It has been observed that the release of water from the Ranganadi Dam into Dikrong has increased erosion and deposition as well as affects the morphology of the river.

Keywords: River morphology · Bank-line shifting · Chanel dimension · Erosion & deposition · GIS

1 Introduction

A river is a complex integrated system of water bodies along with the materials carried with it. The change taking place in different parts of the sphere of the earth every dawn attributes to the alteration of morphological characteristics of the river system. River morphology, in simple words, is the changes in river planforms and their cross section that results from different reasons. The shifting of the channel, width of the channel, erosions, and depositions of river banks is some of the prime parameters to be studied in

R. K. Bhattacharjya et al. (eds.), *Sustainable Water Resources Management*,
Advances in Sustainability Science and Technology,
https://doi.org/10.1007/978-981-16-7535-5_21

a morphological study. There are various hydrological, hydrodynamic, and geological interferences for such alterations.

Recent advancements in aerial photographs, satellite imageries, remote sensing, and GIS technology are currently in vogue for efficient water resources application. It helps gather information from remote areas where field survey is inaccessible, thus standing as a fundamental tool for morphological studies of the rivers.

The river Dikrong carries plenty of water diverted from the Ranganadi hydroelectric power plant, situated in Yazali of Lower Subansiri District of Arunachal Pradesh. The installed capacity of the plant is 405 MW. The project was implemented by North Eastern Electric Power Corporation Limited (NEEPCO) and commenced at the beginning of 2002. The water from the plant is diverted to the Dikrong River through a tunnel. This inter-basin transfer of water might have disturbed the morphology of the river.

The location, operation policies, purpose of the dam, and its size affect the flow and sediment regime, thereby contributing to the morphological changes of a river [9]. These alterations of the flow regime and river morphology have an adverse effect on flood plains with a negative consequence for biodiversity [11]. Thus, the studies on morphological changes due to the construction of dams have gained significance.

Many studies conducted across the world depict the change in the morphological parameters of a river. The lateral movement of the river channel along with the bank erosion and accretion was studied by Lovric and Tosic [8] using remote sensing and GIS at the lower part of the Bosna River during 1958–2013. They concluded that the average movement in the riverbed was 132.4 m with 2.5 m/year lateral channel migration and 10.7074 km^2 bank accretion.

In India, Chauniyal et al. [4] studied the changing pattern of channel morphology on Alaknanda River in Srinagar Valley (Garhwal Himalaya) between 2004 and 2014, using LISS IV data and Google image. They observed that after the 2013 floods, there were drastic changes in river morphology. The comparison of the terrestrial photographs before and after construction of the Supana dam shows that near the point bars and lateral bars, the dam excavated material was deposited. But during the 2013 flood, the material was washed out and affected the channel morphology by increasing the amount of sediments. Again Pal and Pani [12] studied the evolving planform morphology of the middle–lower part of the Ganga River, India, using topographical maps and satellite images. They found that the floodplain experiences autogenic changes and allogenic changes evolution. The allogenic evolution is primarily concerned with the Quaternary evolution of the river.

If we see similar studies by researchers from Assam, there have been many more studies. Phukon and Machahary [13] studied that the bankline shifting of the Jiabharali River from 1967 to 2006 found that there has been a preferential eastward bankline migration in some segments of this river while maintaining the constancy of spatial position in other segments. Sarma [14] studied the fluvial process and morphology of the Brahmaputra River using the IRS satellite image and concluded that erosion and accretion cause bank migration of the Brahmaputra on both banks. Sarma and Phukan [15] carried out a study from 1912–28 to 1963–75 to assess bank erosion and bankline migration of the Brahmaputra River. They found that about 980 km2 area was eroded from 1912 to 1928 whereas only about 112 km^2 was deposited from 1963–75 to 1996 with the overall effect of erosion from 1912 to 1966 was about 868 km^2. Barman [2] studied the fluvial–morphological impact of Pagladiya–Mora-Pagladiya rivers using toposheets and Landsat imageries from 1971 to 2001. He found that Pagladiya and Mora-Pagladiya rivers and their tributaries have changed their courses. The study concluded that the Pagladiya River shows an eastward shifting trend in the past 100 years. Sarma et al. [16] study on change in the channel morphology and bank erosion of the Burhi Dihing River (Assam) in a GIS environment during 1934–1972, 1972–2001, and 2001–2004. They found that during 2001–2004, the average rate of erosion of the river was highest, while during 1972–2001, deposition was highest in both banks. Gogoi and Goswami [5] carried out a study on bankline shifting associated with the erosional pattern of the Subansiri River, Assam, using remote sensing and GIS technology from the year 1915 to 2010 and found that there was about 82.089 km^2 erosion and 43.231 km^2. deposition occurred during the study period. Baishya and Sahariah [1], in their study of bank erosion and bankline shifting of the Baralia River using geospatial technology from 1970 to 2008, found that both the upper and middle parts were highly affected by the erosion problem than in the lower part of the channel. They observed that width increases mostly in the central part due to erosion than in the upper and lower part of the Baralia River. Kotoky and Dutta [7] studied the morphological changes in the Dhansiri River, Assam, between 1914–2000 using GIS and remote sensing techniques. They found that the total average annual erosion and deposition were 1.32 and 1.27 km^2/year, respectively, with the average rate of about 0.006375 and 0.00625 km^2/km erosional and depositional length of the river. Bhuyan and Syiemlieh [3] studied the impact of surplus water from the Ranganadi hydro plant on environmental degradation in the lower Dikrong basin of Assam. They observed that the erosion problem is more acute in the Dikrong basin and seems to be related to nature as well as human influence. On the other hand, the study of morphometric characteristics of the Dikrong River by Huda [6] stated that the upper parts of the catchment create flood problems in the lower reaches due to the basin's geological conditions and morphometric features.

Thus, this study was carried out with the primary objective to evaluate the impact of dam release to Dikrong on different morphological parameters of the river.

Study Area

The river Dikrong starts its journey from the lesser Himalayan ranges in Arunachal Pradesh at an elevation of 2579 m and flows mainly towards the south. Almost parallel to Subansiri, Dikrong came down to the plains crossing NH 15 and railway line from Rangia to Murkongselek on the north bank of Brahmaputra and gets merged with Subansiri River near Badatighat. The river extends between latitude 26°55′N and 27°20′S and longitude 93°15′E and 94°E with a total catchment area of 1395.15 km^2. It flows about 113 kms through the hilly region of Arunachal Pradesh and 32 kms through the plains of Assam [10]. In the present study, a reach of 38.98 km from Nirjuli to the confluence of the Subansiri River is considered (Fig. 1). It is considered based on the confluence point of the diverted water channel of the Ranganadi River to Dikrong River after the dam's construction, which adds plenty of water that might change the river morphology and flow parameters of the river.

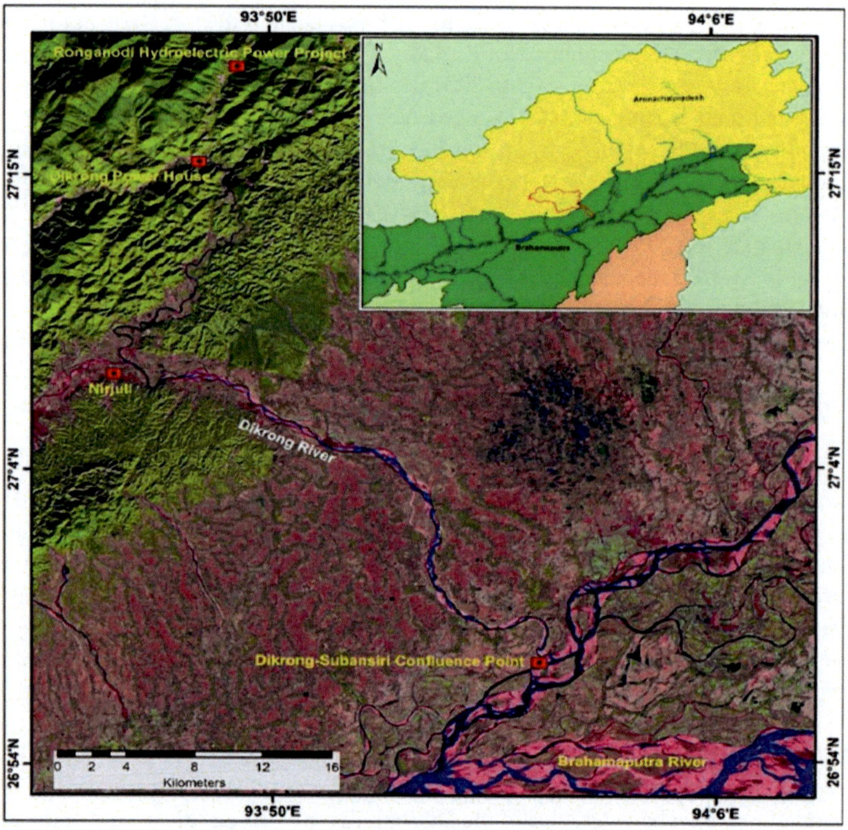

Fig. 1 Geographical location of study area

2 Materials and Method

The present approach employs three phases: data collection, analysis using remote sensing and GIS techniques, and finally, the results.

Data Collection

Initially, the Landsat datasets with a resolution of 30 m of the study area were collected from 1990 to 2019 to investigate spatial changes over time, as shown in Table 1.

Data Analysis

The collected data were composed using Arc GIS to obtain a single raster dataset from the multiple bands. To understand the changing channel course over the study period, i.e. 1990–2019, bankline shapefiles of each year were delineated for the study area. The obtained channels of different years were overlapped to find the change in the pattern of the river course. A comparative quantitative analysis was done for the years 1990, 1995,

Table 1 List of images used in the study and its properties

Data	Resolution (m)	Date	Source
Landsat-4 & 5 TM	30	1990–12-27	USGS Earth Explorer
Landsat-4 & 5 TM	30	1995–12-09	USGS Earth Explorer
Landsat-7 ETM	30	2001–10-22	USGS Earth Explorer
Landsat-7 ETM	30	2005–11-02	USGS Earth Explorer
Landsat-7 ETM	30	2010–12-18	USGS Earth Explorer
Landsat-8 OLI	30	2019–03-14	USGS Earth Explorer

2005, 2010, and 2019 with respect to 2001 as a base year since Ranganadi HEP started operation by this year. All the datasets were divided into 3 km reaches for the studied length of the river to calculate the bankline shifting for the left bank as well as for the right bank of all the five years, as shown in Fig. 2. Again, to study the changes in the dimension of the river, centre lines for each year were drawn, and then, the width of the river at every section for all the study periods was measured. A plot of the width against each section was then plotted in excel.

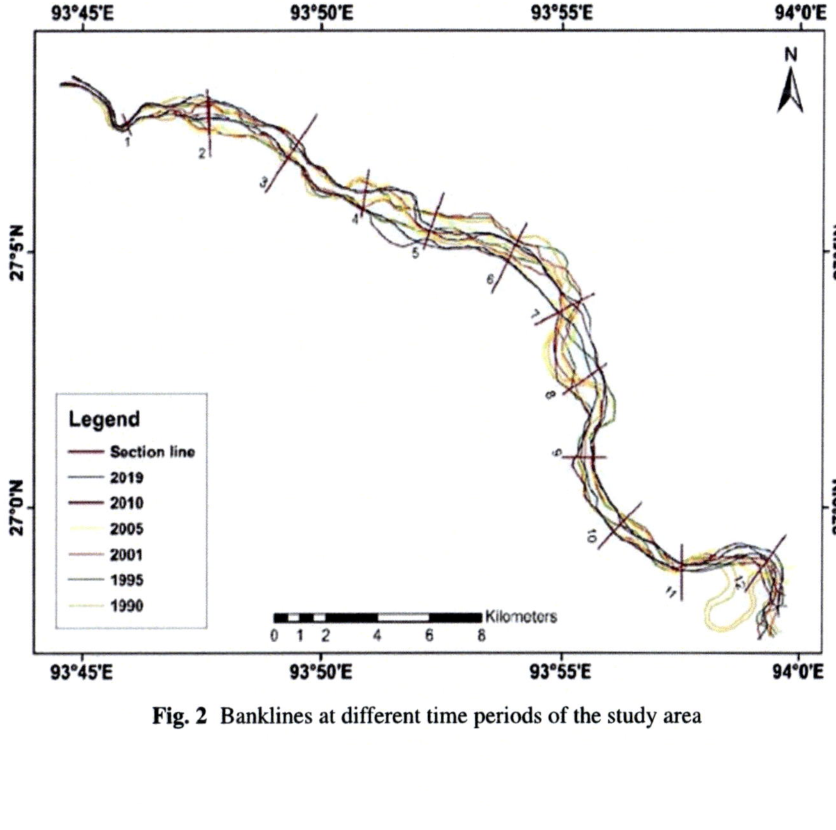

Fig. 2 Banklines at different time periods of the study area

For the further study of erosion and deposition, the river length under study, i.e. 38.98 km, is divided into four equal reaches, as shown in Fig. 3. The overall erosion–deposition phenomenon considering the left bank and right bank was analysed. The right and the left bank of the river under study were digitized manually in ArcGIS10.4. The banklines of different years were then overlaid with respect to the year 2001, and polygon shapefiles are created wherever changes are observed, showing erosion and deposition. The areas of the polygons were calculated to quantify the amount of erosion and deposition along the banks (Fig. 4). In the figure, the red-marked area shows the amount of erosion, and the green-marked area shows the amount of deposition. The total amount of erosion and deposition is then calculated for each reach in Microsoft Excel and plotted accordingly.

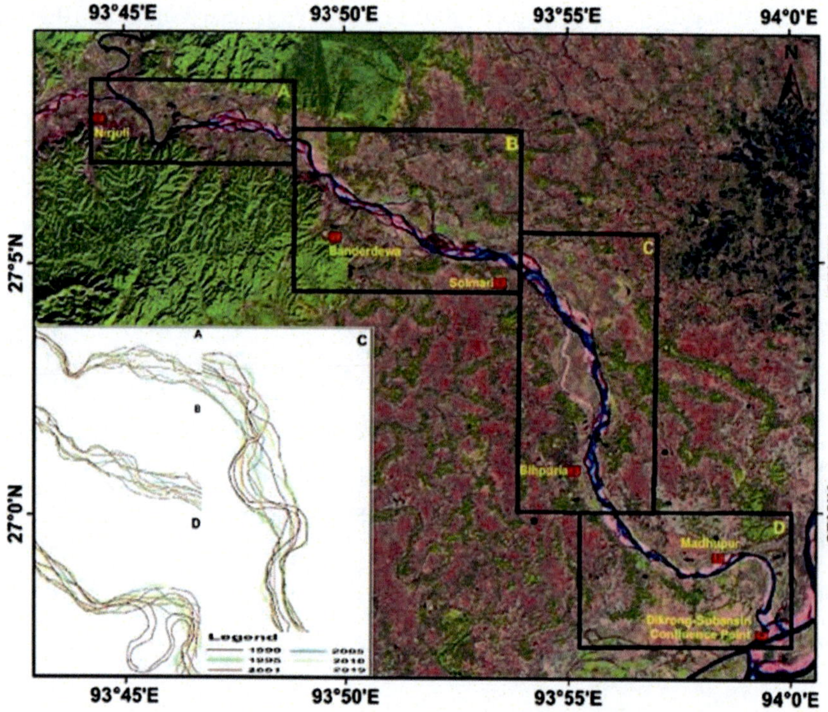

Fig. 3 Reaches for erosion–deposition study of the study area

3 Results and Discussions

Figures 5, 6, 7, 8 and 9 show the shifting of the banklines either northward or southward on both the banks for the different datasets in different time periods. If we see a decade before and after the construction of Ranganadi Dam, during the period of 1990 to 2001, the maximum shift was up to 0.522 km towards northward, and 0.19 km towards southward

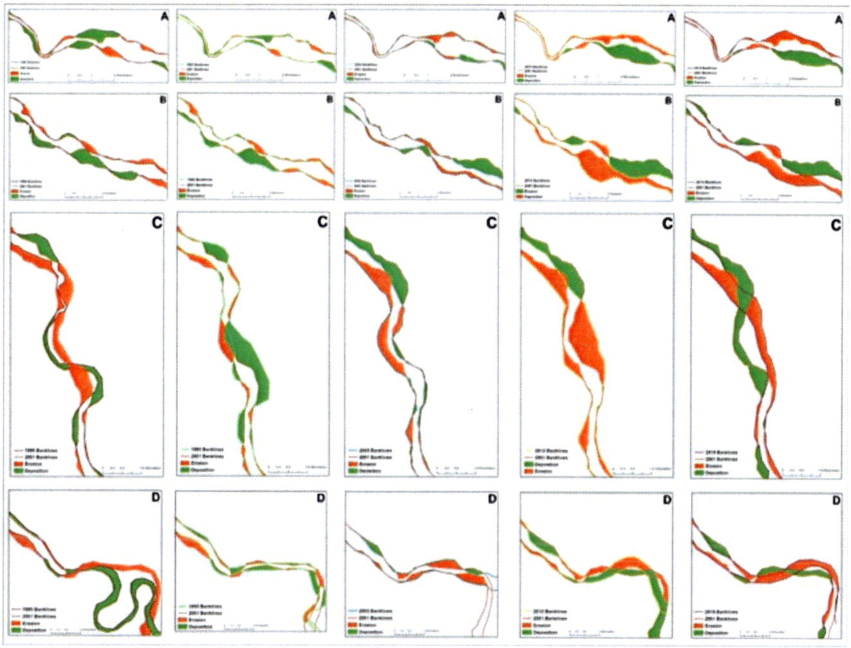

Fig. 4 Amount of erosion and deposition along the banks for each reach

were observed in the left bank. Also, a maximum shift up to 0.464 km towards northward and 0.275 km towards southward was observed in the right bank. On the other hand, from 2001 to 2010, the maximum shift up to 0.774 km towards northward and 0.695 km towards southward was observed in the left bank. Moreover, a maximum shift up to 0.354 km towards southward with 0.596 km towards northward in only one section was observed in the right bank. Thus, the results for 2001–2010 signify that the river got shifted to the southward direction. This might be because of the effect of the water flowing from Ranganadi HEP into the river Dikrong.

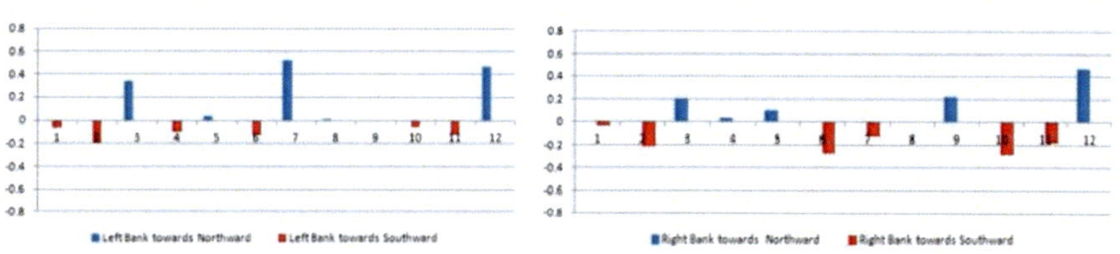

Fig. 5 Bank shifting for the period 1990–2001

The plot of the dimension (width) of the studied reach of the river of each section (Fig. 10) shows an irregular pattern of change in the width of the channel. Where there is an increase in width in one of the banks or sections, there is a decrease in the other bank or section. No significant pattern of increase or decrease in the width is observed primarily due to the dam. However, there is an increase in the river's width in section

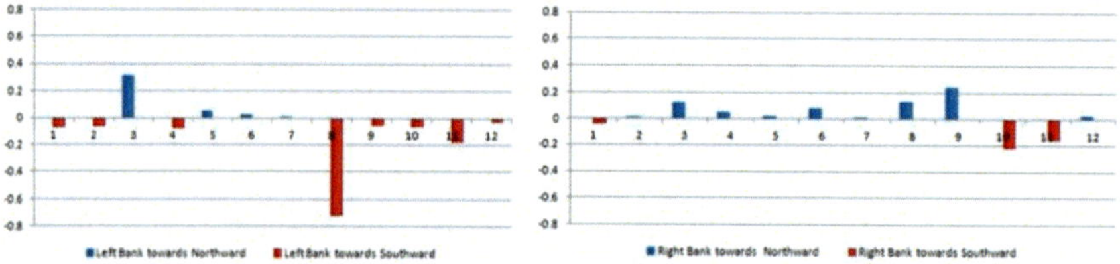

Fig. 6 Bank shifting for the period 1995–2001

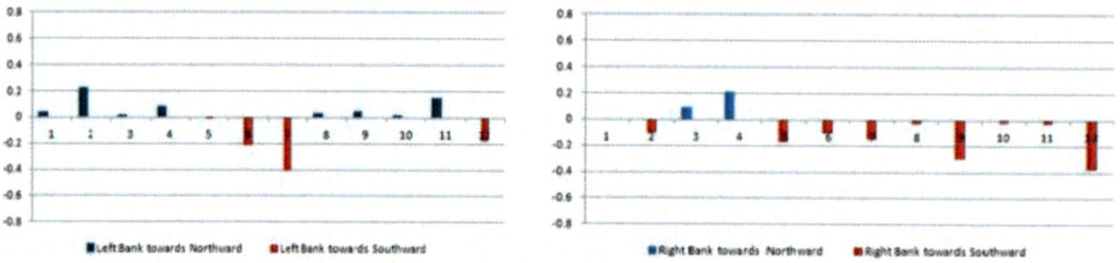

Fig. 7 Bank shifting for the period 2001–2005

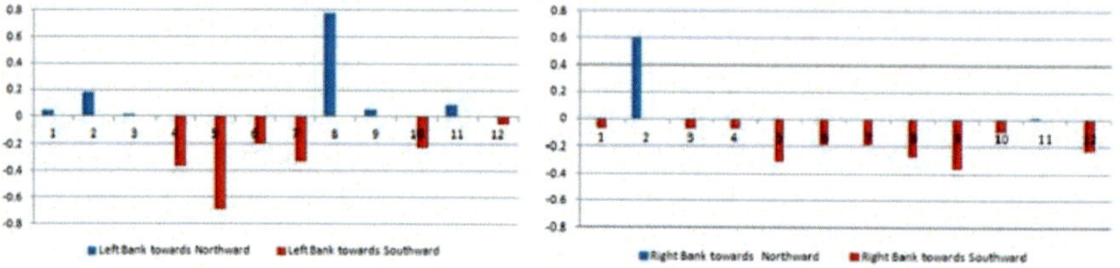

Fig. 8 Bank shifting for the period 2001–2010

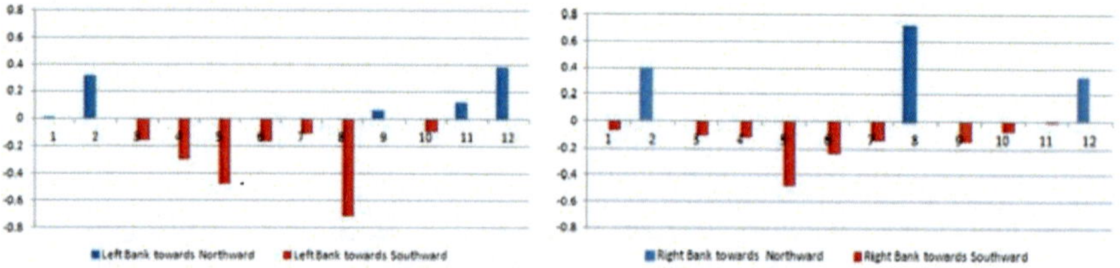

Fig. 9 Bank shifting for the period 2001–2019

number 7 for the year 1990, which might be due to the sharp turn of the river at that portion.

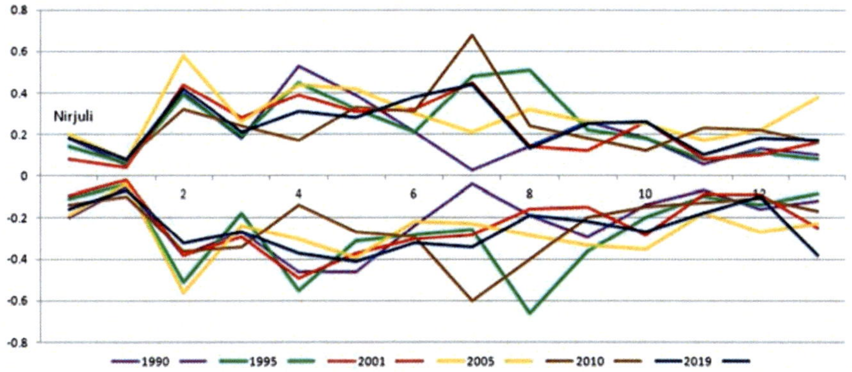

Fig. 10 Plot of channel dimension

The study of erosion and deposition depicted a total erosion of 7.344 km^2 and deposition of 6.789 km^2 occurred over the time period 1990–2001. The erosion and deposition that occurred during 1995–2001, 2001–2005, 2001–2010, and 2001–2019 are 3.909 km^2 and of 5.540 km^2, 4.950 km^2 and 3.566 km^2, 12.130 km^2 and 5.980 km^2, and 9.320 km^2 and 6.770 km^2, respectively.

Figures 11 and 12 show both banks' total erosion and deposition in different time periods for the four reaches. Rate of erosion increased from 0.667 km^2/year in 1990–2001 to 1.3478 km^2/year in 2001–2010. Thus, it can be said that after the construction of the Ranganadi Dam, the excess water that was added to river Dikrong has interfered with the amount of erosion and deposition till 2010. However, till the end of 2019, total erosion got reduced with an increase in deposition rate. Moreover, an interesting result was found that in most cases, if erosion occurred on the left bank, then there is deposition on the right bank or verse versa.

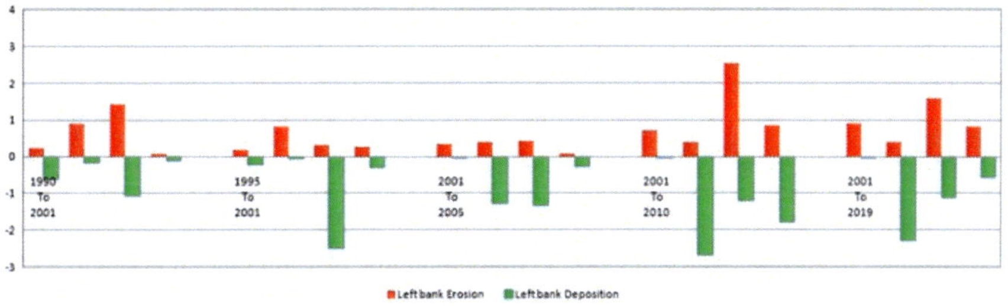

Fig. 11 Amount of erosion and deposition (in sq. km.) along the left bank for each reach in the different time period

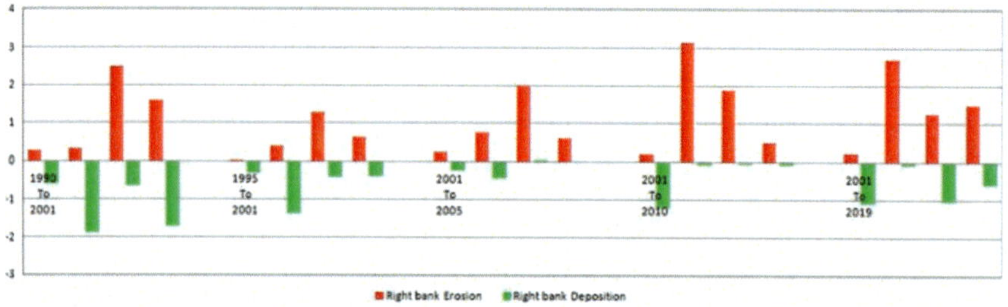

Fig. 12 Amount of erosion and deposition (in sq. km.) along the right bank for each reach in different time periods

4 Conclusions

The present study results reveal that dam operation influences the morphological characteristics of river flow with respect to time and space. There are some specific findings inferred from the present analysis, which are as follows:

- Erosion in the river Dikrong has increased, and deposition has decreased along with shifting of the right bank to the river towards southward, which might be due to the release water of the Ranganadi Dam into the Dikrong.
- The bankline for the decade 2001–2010 signifies that maximum shift up to 0.774 km towards northward and 0.695 km towards southward was observed in the left bank and maximum shift up to 0.354 km towards southward with 0.596 km towards northward in only one section was observed in the right bank.
- The rate of erosion increased from 0.667 km^2/year in 1990–2001 to 1.3478 km^2/year in 2001–2010, which might attribute to the influence of the Ranganadi Dam.
- The plot of the dimension (width) of the studied reach of the river of each section shows an irregular pattern of change in the width of the channel. If there is an increase in width in one of the banks or sections, there is a decrease in the other bank or section. No significant pattern of increase or decrease in the width is observed primarily due to the dam.

References

1. Baishya S, Sahariah D (2015) A study of bank erosion and Bankline migration of the Baralia River, Assam using remote sensing and GIS. Int J Curr Res 7(11):23373–23380
2. Barman S (2007) Fluvio-morphological impact on Pagladiya-Morapagladiya Rivers on rural settlements. PhD thesis, North Eastern Hills University, Shillong
3. Bhuyan G, Syiemlieh HJ (2014) Management of natural resources for sustainable development: challenges and opportunities, vol 9(1). Department of Geography and Resource Management, School of Earth Sciences, Mizoram University, Aizawl, pp 43–53
4. Chauniyal DD, Dutta S, Semwal S (2018) Changing pattern of channel morphology of Alaknanda River in Srinagar Valley (Garhwal Himalaya), India. J Indian Soc Remote Sens 46(9):1379–1387

5. Gogoi C, Goswami DC (2013) Study on bank erosion and bank line migration pattern of the Subansiri River in Assam using remote sensing and GIS technology. Int J Eng Sci (IJES) 2(9):01–06
6. Huda EA (2017) Morphometric characteristics of Dikrong River catchment in the foot-hills of Arunachal Himalayas. IOSR J Human Soc Sci (IOSR-JHSS) 22(7):51–60
7. Kotoky P, Dutta MK (2015) Mechanisms and spatio-temporal variations of meandering and erosion-deposition statistics of the Dhansiri River, Assam. Environ Manage River Basin Ecosyst 1(1):253–282
8. Lovric N, Tosic R (2016) Assessment of bank erosion, accretion and lateral channel migration using remote sensing and GIS: case study—lower part of the Bosna River. Quaestiones Geographicae 35(1):81–92
9. Marren PM, Grove JR, AngusWebb J, Stewardson MJ (2014) The potential for dams to impact lowland meandering river floodplain geomorphology. Sci World J 2014(1):1–24
10. Master Plan Report of Brahmaputra Board, Ministry of Water Resources, Government of India (1996)
11. Mwelwa-Mutekenya E, Crosato A, Wright N, Beevers L (2013) Analysis of the state of the flow and morphology interaction in the hydropower dominated middle Zamdezi subcatchment. In: Proceedings of the 8th international conference on geomorphlogy: geomorphology and sustainability, Groupe Francais de Geomorphologie (GFG), Paris, France
12. Pal R, Pani P (2019) Remote sensing and GIS-based analysis of evolving planform morphology of the middle-lower part of the Ganga River, India. Egypt J Remote Sens Space Sci 22(1):1–10
13. Phukon P, Machahary R (2011) Bankline migration of the Jiabharali River, North Brahmaputra Plain, Assam. Memoir Geo Soc India 77(1):547–557
14. Sarma JN (2005) Fluvial process and morphology of the Brahmaputra River in Assam, India. Geomorphology 70(1):226–256
15. Sarma JN, Phukan MK (2006) Bank erosion and bank migration of Brahmaputra river in Assam during twentieth century. J Geo Soc India 68(1):1023–1036
16. Sarma JN, Borah D, Goswami U (2007) Change in river channel and bank erosion of the Burhi Dihing River (Assam), assessed using Remote sensing data and GIS. J Indian Soc Remote Sens 35(1):93–100

Chapter 22
River Basin Development

Krishna Kamal Das[1] (✉) and Bibhash Sarma[2]

[1] Civil Engineering Department, Bongaigaon Polytechnic, Bongaigaon, Assam, India
[2] Civil Engineering Department, Assam Engineering College, Guwahati, Assam, India

Abstract. For sustainability of any water resources system, the development of river basin is very essential. In this paper, the development of Kulsi River basin has been undertaken for morphometric study and prioritization of sub-watersheds for soil management, irrigation, hydropower generation, and flood mitigation. Morphometric analysis using remote sensing and GIS reveals that the basin is elongated with less runoff and less prone to soil erosion. CROPWAT 8.0 is used for determination of crop water requirements of thirty-nine varieties of crops suitable for the basin. Linear programming model is developed and solved in LINGO for irrigation planning. Three irrigation management plans are suggested for the basin for optimal cropping pattern which gives the maximum net benefits. To investigate the potential of hydropower and irrigation in the basin, a simulation model is developed. From the model, optimal reservoir capacity, firm power, maximum annual power, and maximum annual irrigation are obtained. The basin is 'water surplus' up to the projected year 2050 for meeting all the demands, investigated using the guidelines provided by National Water Development Agency. Finally for flood mitigation, flood retention times have been estimated corresponding to flood absorption capacities at different dam heights.

Keywords: Remote sensing and GIS · Morphometric analysis · CROPWAT · Linear programming · LINGO · Simulation

1 Introduction

The limited water resources in river basins are under tremendous pressure due to population growth and rapid developmental activities. For sustainable use of the limited water resources, development of river basin is a high priority. Gain and Giupponi [1] noted that the water crisis in a river basin is due to improper planning and managements of water resources. Asmamaw [2] stated that proper planning, management, and development of river basin are vital for minimizing the conflicts in water demands in a basin. For the present study, Kulsi River basin is selected. The basin (Fig. 1) is located between 25°35′N & 26°07′N and 90°45′E & 91°00′E. Total catchment area of the basin is 3770 km^2 which occupies in Kamrup and Goalpara district of Assam as well as West Khasi hills and East Garo hills district of Meghalaya. The Kulsi multi-purpose project

R. K. Bhattacharjya et al. (eds.), *Sustainable Water Resources Management*,
Advances in Sustainability Science and Technology,
https://doi.org/10.1007/978-981-16-7535-5_22

is proposed jointly by the Brahmaputra Board, Central Water Commission, and Government of Assam for hydropower generation, irrigation, and flood mitigation. For the development of Kulsi River basin, firstly morphological parameters of the basin are studied to recognize the drainage characteristic and prioritization of sub-watershed for land management. Secondly, crop water requirements for the basin have been estimated to ascertain the total irrigation requirements for the basin. Thirdly, irrigation planning has been done to find the optimal cropping pattern to maximize the net benefit. Fourthly, the optimal reservoir capacity, firm power, maximum hydropower generation, and maximum irrigation that can be possible in the basin are studied by development of simulation model applicable to reservoir planning.

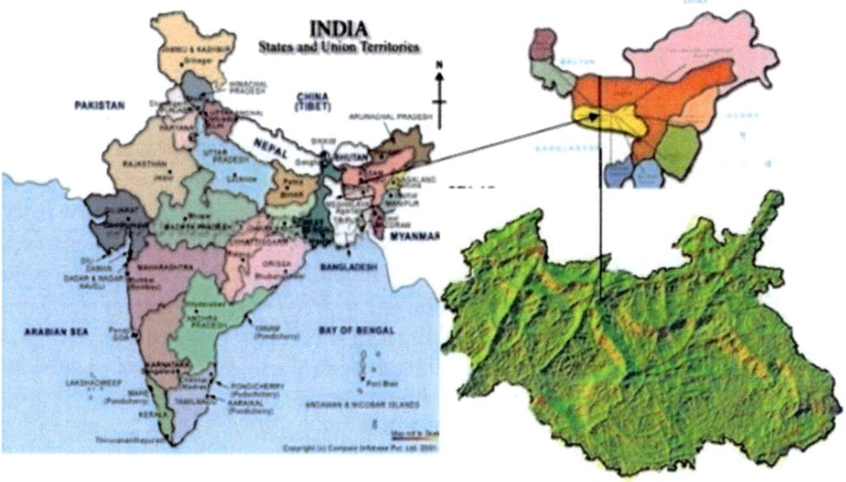

Fig. 1 Location map

2 Methodology

River basin development involves careful planning and management of various complex tasks. For developmental activities of the basin, following methodologies are adopted.

For morphological study about a basin, remote sensing and GIS are modern tools. Sangita Mishra and Nagarajan [3] explained the importance of GIS for analyzing the morphological features and prioritization of sub-watersheds for managing land in a basin. Horton [4] and Strahler [5] established a number of morphometric parameters and their relationships to study the basin morphology. ArcGIS 10.1 software developed by ESRI has been used for morphological analysis of the basin.

For planning and development of irrigation system in a basin, accurate estimation of crop water requirement is very essential. Bouraima et al. [6] noted that for optimum use of limited water, a clear knowledge is essential on crop water requirements. Abdelhadi et al. [7] compared the method developed by Farbrother and Penman and found that Penman method was more accurate for assessment of total crop water requirements. CROPWAT 8.0 software based on Penman–Monteith method, developed by Food and Agriculture Organization (FAO), has been used for determination of crop water requirements of various crops suitable for the basin.

In water resources management practices, the linear programming model is widely used for irrigation planning. Shrivastava et al. [8] stated that optimization model is very essential for obtaining benefit and optimum cropping pattern for planning and management of water resources systems. Chowdhury and Chakrabarty [9] stated that solving the LP model in LINGO solver is promising for allocation of areas to maximize crop production and optimal use of irrigation water. LP model is widely used because it can accommodate large numbers of decision variables and constraints. Here, LP model is developed for maximizing the net benefit considering the constraints such as water availability, land availability, protein, and calorie requirements for the basin population for the projected year 2050 and solved in LINGO solver.

For finding storage capacity, firm power, irrigation, and hydropower potential, and their reliabilities for the development and management of river basin, a simulation model has great importance. Bosona and Gebresenbet [10] developed a simulation model to control the reservoir releases and stated that a small advancement in reservoir operation leads to a large benefit. Lin and Rutten [11] stated that management of a reservoir system is very complex, and this could be overcome by optimal operation of the reservoir system by using optimization and simulation models. In this paper, simulation model is developed for sizing of the reservoir, study of hydropower, and irrigation potential in the basin without compromising target reliabilities.

Accurate assessments of status of water either surplus or deficit in a river basin are very important for any developmental activities in the basin. Juniatil et al. [12] noted that for effective land use, water availability is an important factor for sustainable water resources. Usmanov et al. [13] stated that water balance is directly related to the agricultural activity in a basin. The guideline provided by the National Water Development Agency (NWDA) is an appropriate method for water balance study in a river basin.

For planning and development of any river basin, management of flood is of utmost importance. Mujere [14] concluded that the Gumbel's distribution method was appropriate for frequency modeling for estimating the amount of flood for a return period. Hettiarachchi [15] noted that the detention reservoirs were useful for flood control measures. Here, Gumbel's extreme value distribution method for estimation of maximum design floods at various return periods is applied, and flood absorption capacities and detention time at different elevations are determined to minimize the impact of flood.

3 Results and Discussions

The morphometric analysis of the basin is done using remote sensing and GIS tool. ArcGIS 10.1 software is used for the analysis of DEM (30 m resolution) and watershed delineation. From the morphometric analysis of Kulsi basin, it is found that the basin is a sixth-order dendritic drainage pattern. The slope ranges from nearly level surface to very steep. The basin is elongated in shape, having less surface runoff and less prone to soil erosion. The soil characteristics of the basin are coarse, hence more permeable.

For management of land on priority basis, prioritization of sub-watersheds is done. The watershed of the basin is divided into seven sub-watersheds, and morphological analyses are done for each sub-watershed. The compound parameter values are calculated and observed that out of seven sub-watersheds of the basin, the watershed SWS 1 and SWS 6 falls in high priority; SWS 3, SWS 5, SWS 7, and SWS 4 are in medium priority, and SWS 2 falls under low priority (Table 1). Since SWS 1 belongs to lowest compound parameter value, so SWS 1 belongs to the highest erosion-prone sub-watershed. The soil conservation measures should be undertaken first in SWS 1 and then to the other sub-watersheds based on the priority. The crop water requirement (CWR) for the basin is calculated using CROPWAT 8.0 software developed by FAO, which is based on Penman–Monteith method for estimation of crop evapotranspiration. Altogether thirty-nine verities of crops suitable for the basin are proposed by the project authority. The crop data and soil and climatic data are incorporated in CROPWAT as recommended by FAO. Monthly crop water requirements are shown in Table 2. Total crop water requirement is estimated to be 3593.2 mm.

Table 1 Priority table

Sub-watershed	Compound parameter value	Priority	Degree of erodibility
SWS 1	2.9	First	High
SWS 2	4.9	Seventh	Low
SWS 3	3.6	Third	Medium
SWS 4	4.0	Fourth	Medium
SWS 5	4.2	Fifth	Medium
SWS 6	3.2	Second	High
SWS 7	4.5	Sixth	Medium

Table 2 Monthly crop water requirements

Months	CWR (mm)	Months	CWR (mm)
Jan	627.5	Jul	0
Feb	536.3	Aug	0
Mar	844.6	Sep	0
Apr	281.7	Oct	113.1
May	43.4	Nov	430.3
Jun	0	Dec	716.3

To obtain best cropping pattern suitable for the basin, which gives the maximum net benefit from agriculture, LP model is developed and solved in LINGO solver. The LP model is developed with the objective function for maximizing net benefits subject to water and land area availability, protein, and calorie requirements for the basin population for the projected year 2050, minimum crop area to protect farmer's affinity toward crops and maximum crop area depending on market limitations. From the results of the model, three management plans for the year 2030 (Plan-1), 2040 (Plan-2), and 2050 (Plan-3) are obtained. The net benefits that can be obtained from Plan-1, Plan-2, and Plan-3 are Rs. 63,915.58 Lakh, Rs. 41,676.38 lakh, and Rs. 10,720.92 lakh, respectively. The total irrigation requirements for the area and crops for the Plan-1, Plan-2, and Plan-3 are 93.282 Mcm, 97.069 Mcm, and 105.825 Mcm, respectively (Fig. 2).

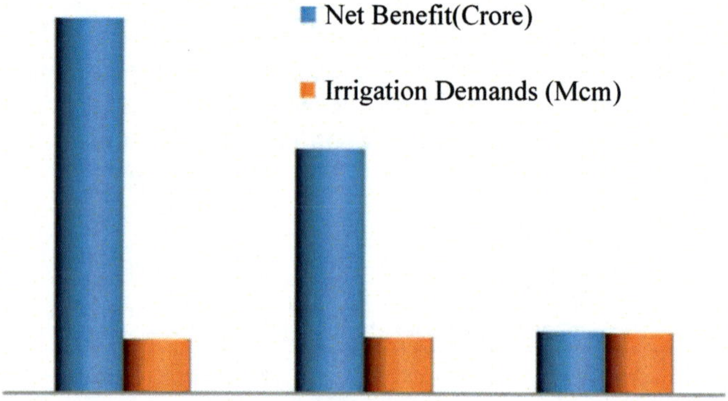

Fig. 2 Net benefits and irrigation demands

A simulation model applicable to reservoir operation is developed to study the irrigation and hydropower potential of the project. Irrigation, hydropower, and ecological demands are considered in the model. The model is run for the irrigation demands as obtained for Plan-1, Plan-2, and Plan-3 for the target of irrigation reliability ≥ 75% and power reliability ≥ 90%. From the project 10.70 MW firm hydropower, 252.292 MW annual power and 155.84 MW average annual power can be produced with overall power reliability 90.14%. The optimal reservoir capacity required is 522.722 Mcm. Average annual irrigation of 91.71, 95.50, and 104.24 Mcm can be done with irrigation reliabilities 94.87%, 94.87%, and 94.66%, respectively, from Plan-1, Plan-2, and Plan-3. Irrigation releases can be increased to 146.502 Mcm, 152.449 Mcm, and 166.201 Mcm, respectively, from Plan-1, Plan-2, and Plan-3 without compromizing the target reliability.

The surface water balance study of the basin is performed as per the guidelines provided by the NWDA. The water requirements for irrigation, domestic, industry, and hydropower are included in the study. The available surface water at 75 and 50% dependability is considered in the study. The water balance study (Table 3) reveals that the basin is water surplus of 467.955 Mcm and 603.585 Mcm, respectively, at 75 and 50% dependability.

Finally, flood retention time at different reservoir level is estimated as a flood control tool for management of flood. Goodness of fit test for Gumbel distribution has been examined using Chi Square. The annual and monthly design floods for the month of May, June, July, August, and September are estimated for return period 1000 year, 100 year, 50 year, 25 year, 20 year, 15 year, 10 year, and 5 year using Gumbel's extreme value

Table 3 Surface water balance (NWDA)

Dependability		At 75% (Mcm)	At 50% (Mcm)
Availability (+)	Surface water	659.34	794.97
Demands (−)	Irrigation	105.825	105.825
	Domestic	111.29	111.29
	Industrial	239.47	239.47
	Hydropower	25.990	25.990
Regeneration (+)	Irrigation	10.58	10.58
	Domestic	89.03	89.03
	Industrial	191.58	191.58
Surface water availability		**+467.955**	**+603.585**

distribution method. The maximum design floods in July are 743.545 Mcm, 528.552 Mcm, 463.487 Mcm, 397.938 Mcm, 376.660 Mcm, 349.035 Mcm, 309.581 Mcm, and 239.651 Mcm, respectively, for return period 1000 year, 100 year, 50 year, 25 year, 20 year, 15 year, 10 year, and 5 year, respectively. The flood absorption capacities from 100.0 to 115.0 m elevation are estimated for every 1.0 m elevation interval, and accordingly, detention time for the absorbed flood is also estimated using simulation (Table 4). The discharge carrying capacity of Kulsi River is also checked for carrying maximum design flood (Fig. 3), and it is seen that the discharge carrying capacity of Kulsi River is far more than the maximum design flood. So, it may be concluded that discharge from Kulsi River is not causing any flood in its upper reaches. But in the lower reaches drainage congestion takes place due to flatter bed slope, joining of discharges from other north flowing rivers, flood of Brahmaputra River. The drainage congestion in the lower reaches of Kulsi River can be minimized by removal of silt and sand from the river bed.

Table 4 Detention time–flood absorption–elevation (July)

Return period (year)		1000	100	50	25	20	15	10	5
Design flood (Mcm/h)		1.033	0.734	0.644	0.553	0.523	0.485	0.430	0.333
Elevation (m)	Flood absorption (Mcm)	Detention time (h)							
115.00	0.000	0.00	0	0	0	0	0	0	0
110.00	98.613	95.49	134.33	153.19	178.42	188.50	203.42	229.35	296.27
105.00	183.750	177.93	250.31	285.44	332.46	351.25	379.04	427.35	552.05
100.00	255.413	247.32	347.93	396.77	462.12	488.23	526.87	594.02	767.35

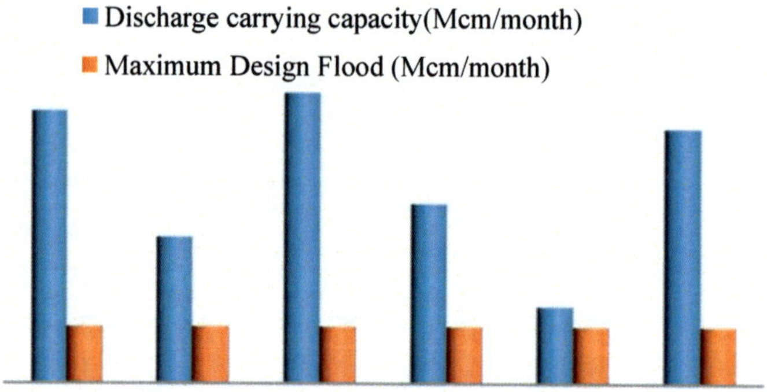

Fig. 3 Discharge carrying capacity of Kulsi River

4 Conclusion

The Kulsi multi-purpose project is initiated by Brahmaputra Board for hydropower generation, irrigation, and flood mitigation. The project is in the inception level. So, it has huge potential for the study. In this study, all the intended purpose of the project is carried out systematically. As such if implemented, the project will benefit the society as well as the country.

References

1. Gain AK, Giupponi C (2015) A dynamic assessment of water scarcity risk in the Lower Brahmaputra River Basin: an integrated approach. Ecol Indic 48:120–131
2. Asmamaw DK (2015) A critical review of integrated river basin management in the upper Blue Nile river basin: the case of Ethiopia. Int J River Basin Manage 13(4):429–442
3. Sangita Mishra S, Nagarajan R (2010) Morphometric analysis and prioritization of subwatersheds using GIS and remote sensing techniques: a case study of Odisha, India. Int J Geomatics Geosci 1(3)
4. Horton RE (1945) Erosional development of streams and their drainage basins: hydrophysical approach to quantitative morphology. Geol Soc Amer Bull 56:275–370
5. Strahler AN (1964) Quantitative geomorphology of drainage basin and channel networks. In: Chow VT (ed) Handbook of applied hydrology. McGraw Hill Book, New York, pp 4–76
6. Bouraima K, Zhang W, Wei C (2015) Irrigation water requirements of rice using Cropwat model in Northern Benin Abdel. Int J Agric Biol Eng Beijing 8(2):58–64
7. Abdelhadi AW, Hata T, HaruyaTanakamaru AkioTada, Tariq MA (2000) Estimation of crop water requirements in arid region using Penman-Monteith Equation with derived crop coefficients: a case study on Acala cotton in Sudan Gezira irrigated scheme. Agric Water Manag 45(2):203–214
8. Shrivastava SK, Verma MK, Devatha CP (2012) Optimization modelling for crop planning of hasdeo bango command. Int J Eng Res Technol (IJERT) 1(9)
9. Chowdhury MA, Chakrabarty D (2015) Optimal crop yield under limited water availability—a linear programming approach. J Basic Appl Eng Res 2(10):892–895

10. Bosona TG, Gebresenbet G (2010) Modeling hydropower plant system to improve its reservoir operation. Int J Water Resour Environ Eng 2(4):87–94
11. Lin NM, Rutten M (2016) Optimal operation of a network of multi-purpose reservoir: a review. Proc Eng 154:1376–1384 (12th international conference on hydroinformatics, HIC)
12. Juniatil AT, Sutjiningsih D, Soeryantono H, Kusratmoko E (2018) Proposing water balance method for water availability estimation in Indonesian regional spatial planning. In: The 4th international seminar on sustainable urban development IOP publishing IOP conference series: earth and environmental science 106/012073. https://doi.org/10.1088/1755-1315/106/1/012073
13. Usmanov S, Mitani Y, Kusuda T (2016) An integrated hydrological model for water balance estimation in the Chirchik River Basin, Northern Uzbekistan. Comput Water Energy Environ Eng 05(03):11. Article ID:68208
14. Mujere N (2011) Flood frequency analysis using the Gumbel's distribution. Int J Comput Sci Eng (IJCSE) 3(7). ISSN: 0975-3397
15. Hettiarachchi (2011) Detention reservoirs as a flood control measures. Engineers: Inst Eng, Sri Lanka XXXXIV(02): 957–60

Water Quality and Management

Chapter 23
Simulation of Fluoride Migration in Groundwater of the Affected Areas of Shilabati Riverbank, West Bengal, India

Arghya Ghosh[1](✉), Suresh A. Kartha[1], and Sandip Mondal[2]

[1] Department of Civil Engineering, Indian Institute of Technology Guwahati, North Amingaon, Guwahati, Assam 781039, India
{arghy,kartha}@iitg.ac.in

[2] Department of Earth and Environmental Studies, National Institute of Technology Durgapur, Durgapur, West Bengal 713209, India

Abstract. To obtain insightful knowledge about the migration pathway of fluoride and the transport phenomenon in the subsurface environment, a three-dimensional conceptual contaminant fate and transport model was developed using numerical groundwater modelling software (Visual MODFLOW 6.0). The conceptual model relied on the measured hydro-geological data and boundary conditions related to hydraulic conductivity, storage coefficient, specific yield, general head, recharge and river boundary. The model calibration yields a linear correlation coefficient of 0.978 for flow model and 0.947 for the transport model. Sensitivity analysis of the calibrated model shows that migration of fluoride is more susceptible to changes in hydraulic conductivity than other input parameters. From the scenario analysis of the conceptual model, different rates of fluoride migration in different stratigraphic layers of the aquifer have been observed. The simulation results inferred that horizontal migration of fluoride had made the surroundings of the contaminated aquifer unsafe for drinking water and irrigation purposes. From both scenario and sensitivity analyses, it has been found that some common hydrologic phenomenon prevails in subsurface environment of the study area that leads to enrichment and migration of fluoride from the contaminated zone to the other part of the model domain along with groundwater flow.

Keywords: Fluoride · Migration · Conceptual model · Contaminant transport

1 Introduction

Groundwater is one of the most bountiful and valuable natural resources of freshwater that demand special consideration. The existence of groundwater beneath the surface of the earth and its flowing capability was always a curiosity to the mankind and from very ancient period onwards human societies investigated on exploiting this resource. However, the overexploitation and improper management of this freshwater reserve have

R. K. Bhattacharjya et al. (eds.), *Sustainable Water Resources Management*,
Advances in Sustainability Science and Technology,
https://doi.org/10.1007/978-981-16-7535-5_24

resulted in declination as well as contamination of groundwater resources. Therefore, knowledge of the behaviours of an aquifer system under different stresses is essential for the protection and management of subsurface groundwater. In this context, solute transport models are an excellent tool for risk assessment and anticipating the future responses of complex aquifer systems and heterogeneous geological formations.

Fluoride is the13th most abundant element in the earth's crust and present in trace amounts in almost all groundwater samples throughout the world [1]. In India, the occurrence of fluoride in groundwater is mainly due to the geological, hydrological and hydrochemical aspect [2]. As rainwater percolates through the vadose zone and reaches the water table, it partly dissolves certain components of fluorine-bearing minerals (such as fluorite, apatite, micas, muscovite and amphiboles.) present in the bedrock, leading to enrichment of fluoride in groundwater [3]. Groundwater chemistry enhancing the fluoride concentration includes pH, the concentration of HCO_3^-, Ca^{2+} and Na^+ ions [4].

Consumption of fluoride higher than the permissible limit (>1.5 mg/L) is the primary reason behind dental and skeletal fluorosis [5]. In India, about 62 million people, which include 6 million children, are severely affected due to the use of fluoride-contaminated water for drinking purposes (Andezhath and Ghosh 2000) [6]. In West Bengal, during the habitation survey conducted by West Bengal Public Health Engineering Department (WPHED) in the year 2003 and subsequently rapid assessment survey conducted by the Fluoride Task Force, West Bengal, in the year 2005, it was found that predominance of excess fluoride in 729 habitations spread over 45 blocks in nine districts [7]. As there was a spreading of fluoride, which increased over time, several types of research were conducted to understand its occurrence and spatiotemporal variation. However, most of these studies were based on the geochemistry of water and did not thoroughly investigate whether the fluoride in new areas of the groundwater domain was due to the migration of fluoride from contaminated pockets. Henceforth, it is important to quantify the migration pathway of fluoride and the transport phenomenon in the subsurface environment.

Although several computational models are available to simulate the phenomenon of this spatiotemporal change in solute concentration in groundwater among them, MT3DMS model is easily accessible, user-friendly and versatile [8]. It has been successfully used in a variety of environment for the different contaminant. Saghravani et al. [9] conducted a study to simulate the migration path of phosphorus in groundwater by MT3DMS for 10 and 50 years period. Ghoraba et al. [10] demonstrated the application of MODFLOW and MT3DMS model to investigate groundwater flow and ammonium transport in the central part of the Nile Delta (El-Gharbiya Governorate). Reed et al. [11] developed a three-dimensional contaminant fate and transport model by utilizing MT3DMS to simulate boron transport in the Upper Florida Aquifer.

Therefore, the objective of the present research is to develop a conceptual model to simulate the transport of fluoride in the groundwater and to observe its change in concentration with time. To meet this objective, fluoride-contaminated groundwater domain of Shilabati riverbank of Bankura district, West Bengal, has been chosen for fluoride migration study.

2 Description of Study Area

Location and Hydrometeorology

The study area is the middle part of Shilabati River Basin in Simlapal block of Bankura district of West Bengal State. It is geographically extended from 22°59′34″ North to 22°54′37.00″ North latitude and 86°58′51″ East to 87° 5′00″ East longitudes. The entire study area is about 120 km^2 with an average elevation of 80 m from mean sea level. The climatic condition of the area under study is tropical with an average annual rainfall of 1400 mm in a year, and 80% of the total precipitation is received from July to October. The maximum temperature reaches at 46 °C during summer and lowers to a minimum of 8 °C during winter. Spatial distribution of fluoride in the study area is clustered and shown in Fig. 1.

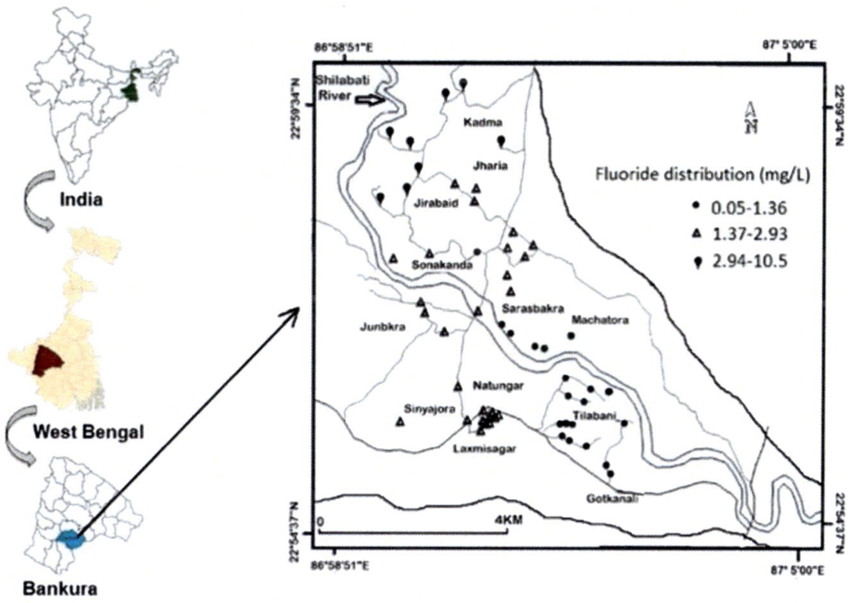

Fig. 1 Study area with groundwater sampling locations

Geology and Hydrology of the Study Area

Geomorphologically, Bankura district is divided into three parts. Hilly hard rock dominates the western part, while the central part is dominated by a mixed formation of both alluvial and rocky segments. The eastern part is a sedimentary alluvial plain, where the present study has been conducted. The lithological diagram of the study area (Fig. 2) shows laterite, and fine loamy soil prevails in the uppermost layer followed by unconsolidated sandy sediments of recent alluvium in the middle layer. The Pre-Cambrian metamorphic layer of granite genesis forms the bottommost layer which contains fluoride-bearing minerals. Groundwater mainly occurs in unconfined condition within the middle and bottommost layer.

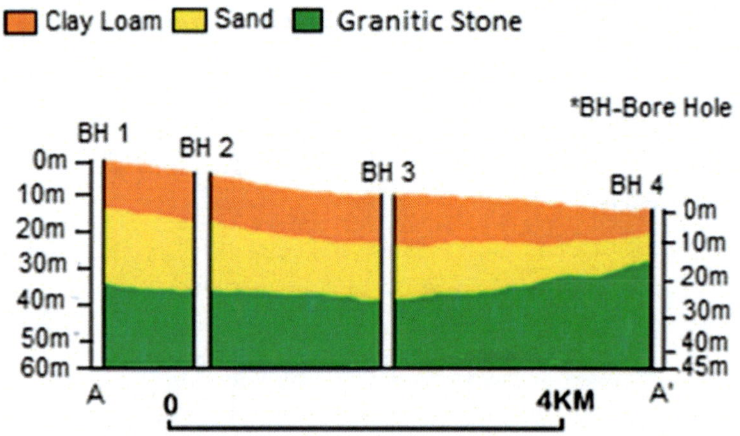

Fig. 2 Subsurface lithological diagram (section AA′ of Fig. 1)

3 Materials and Methods

Determination of Model Input Parameters

In the present study, two categories of input parameters have been assigned to the model. The first category includes the various parameters to describe hydrological and geochemical conditions in the model domain. The other category comprises flow rate, hydraulic head, contaminant concentration and its arrival time, etc.

Hydraulic Conductivity

To determine the hydraulic conductivity, soil samples collected from different stratigraphic layers are subjected to sieve analysis. Allen Hazen equation was used to estimate the hydraulic conductivity of soil samples [12].

$$K = 0.01d_{10}^2, \tag{1}$$

where K = hydraulic conductivity, and d_{10} = effective size of the soil particles (mm), determined from the grading curve.

The horizontal hydraulic conductivities of clayey loam, sand and fractured granitic stone are estimated as 0.5 m/day, 20 m/day and 2 m/day, respectively. Vertical hydraulic conductivities are assumed as 10% of the horizontal hydraulic conductivities [13].

Porosity

Porosity affects transport calculation in two important ways. One way is that this is a parameter to determine the seepage velocity that controls the advective transport, and in another way, it is used advective transport, and in another way, it is used to determine the volume of pore of a cell available for storage of solute mass. In the governing equation of solute transport, only effective porosity is considered, which is smaller than total porosity which reflects the presence of immobile water in the pore space with zero groundwater seepage velocity. In the present study, typical values of 0.15 and 0.3 for effective and total porosity have been assumed for modelling purpose [14].

Dispersivity

It is difficult to define dispersivity values for field-scale contaminant transport problem. Numerous studies have been undertaken to estimate field dispersivity values. Values of dispersivity generally appear to be dependent on the scale of testing or observation. Xu and Eckstein suggested one empirical formula to calculate longitudinal dispersivity α_L, which is given as [15]:

$$\alpha_L = 0.83(\text{Log}\, L)^{2.414} \tag{2}$$

where L is the length of the cell.

Since the length of a cell in the model domain is 187 m, the longitudinal dispersivity has been calculated as 6.017 m. Transverse dispersivity is generally observed to be smaller than longitudinal dispersivity and considered as 10% for horizontal transverse dispersivity and 1% for vertical transverse dispersivity [16].

Development of Conceptual Model

To develop the conceptual model, the entire study area has been discretized into 40 rows (x-axis) 50 columns (y-axis) and two layers (z-axis). Each layer of the model contains 2000 cells, and each cell covers an area of about 0.06 km^2. A uniform thickness of 20 m has also been assumed for each layer to minimize the discretization error. During the simulation, cells covering the area having maximum fluoride concentration (10 mg/L) have been considered as the source of the contaminant in the modelling domain. The conceptual model, along with different boundaries, is shown in Fig. 3.

Model Boundaries

Three general boundary conditions have been assigned to the model for simulating migration of fluoride due to advective and dispersive transport. The boundary conditions are described as follows:

Recharge Boundary

To estimate the groundwater recharge from rainfall, ten years rainfall data (2007–2017) were used, and the recharge was calculated using the following empirical formula [17]:

$$R = 3.984(R_{\text{av}} - 40.64)^{0.5}, \tag{3}$$

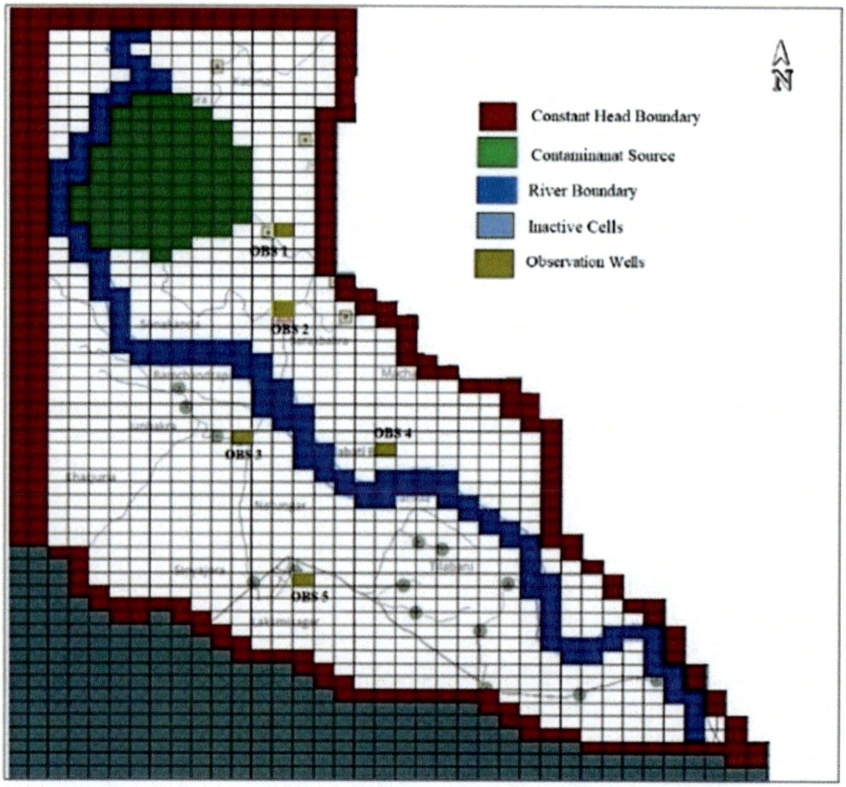

Fig. 3 Conceptual model with various boundaries and observation wells

where R= areal recharge (cm), and R_{av}= the average annual rainfall (cm). As per the guidelines suggested by the Groundwater Estimation Committee (1984), the following provisions are taken during the calculation [18].

(a) Recharge from irrigated rice area is about 35–64% of tube well discharge, and
(b) Recharge from irrigated non-rice area is about 25–36% of tube well discharge.

Based on the calculation, a total recharge of 150 mm/year has been assigned on the uppermost layer of the model domain.

Constant Head Boundary

Constant head boundary (Fig. 3) has been assigned to the model for both the layers. For the northern part, it is 100 m, and for the southeastern part, it is 60 m. Heads are calculated based on the depth of the groundwater table and the height from mean sea level (MSL). The direction of groundwater flow is from northwest (NW) to southeast (SE) following the slope of the land surface.

River Boundary

A river boundary (Fig. 3) has been applied in the uppermost layer of the model. River boundary allows water to enter or leave the model. To set river boundary condition along Shilabati River, the river stage and riverbed bottom are taken as 49.5 m and 48.5 m, respectively [19]. The average width of the river is taken as 250 m.

4 Result and Discussion

Model Calibration

Model calibration is a process that involves the adjustment of model input parameters so that the resultant predictions satisfactorily match a set of targets [20]. Then, this calibrated model can be used for scenario analysis or predicting future responses. However, it is difficult to do so because the exact location of contaminant sources is often uncertain. Traditionally, calibration of the model is conducted by the trial-and-error method, measures of goodness-of-fit, etc. Model fitness is generally evaluated by visually comparing the observed and simulated values or by error-based weighting which includes mean residual error, absolute mean residual error, linear correlation coefficient and root mean squared error (RMSE) [21]. During calibration, it is crucial to decide whether the calibration to be steady state (matching field observations), transient (matching a time series) or both. The classical conceptual framework of the temporal evolution of a solute plume involves transient simulations of solute transport, assuming that the groundwater flow field is under steady-state condition [22].

The scatter plots (1:1 plot) for flow and transport model calibration are presented in Fig. 4. From the plots, it can be observed that for flow model, the correlation coefficient and RMSE are 0.978 and 1.37 m, respectively, and for the transport model, these values are 0.947 and 0.398 mg/L, respectively. Based on the previous studies of field-scale solute transport modelling, it can be inferred that the obtained values are significant and within acceptable limits.

Scenario Analysis

To predict migration of fluoride in the study area and change in concentration with time, calibrated model was run for various stresses on the aquifer, and two scenarios were assessed. As the diffusion of fluoride in the soil is very low (0.044 m^2/year), long simulation period has been taken. The first scenario (Fig. 5a, b) of ten years simulation period shows that spreading of fluoride is not substantial, and the concentration of fluoride is below the permissible limit in most of the areas. In the second scenario (Fig. 5c, d) of twenty years simulation period, it is found that almost entire modelling domain has been contaminated with fluoride. It is also observed that migration of fluoride is different for two layers which may be due to the difference in hydraulic conductivity of the layers.

Sensitivity Analysis

To investigate the stability of the calibrated model due to changes in model input parameters and hydrological stress, sensitivity analysis has been performed. In the present

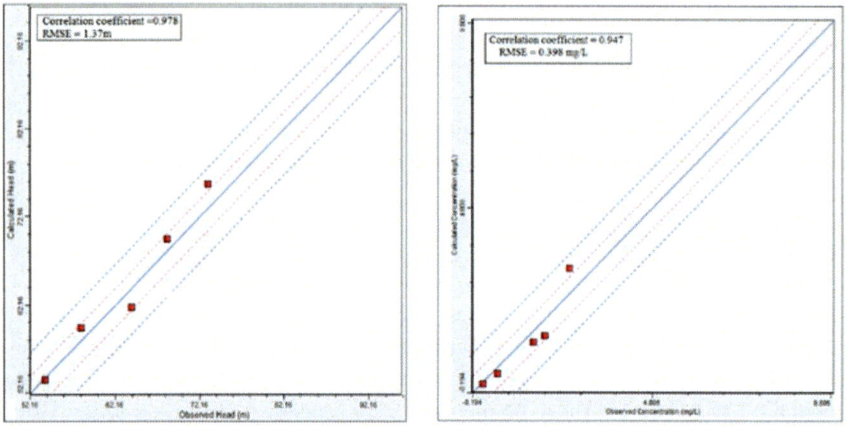

(a) (b)

Fig. 4 Scatter plot of **a** observed versus calculated head and **b** observed versus calculated concentration

study, 50% increase and decrease in the values of the general head, hydraulic conductivities and recharge have been assigned to check the sensitivity of the calibrated model [23]. The plots of RMSE versus the calibrated parameter/input values are shown in Fig. 6. From these plots, it can be observed that fluoride migration is the most sensitive to changes in hydraulic conductivity than other parameters.

5 Conclusion

In the present research, migration of fluoride in groundwater and its change in concentration with time have been assessed by developing a contaminant transport model using MT3DMS. Transient state simulation of the calibrated model reveals different rates of fluoride migration in different stratigraphic layers of the aquifer, which may be due to the difference in hydraulic conductivity of the layers. Sensitivity analysis of the calibrated model shows that migration of fluoride is more susceptible to changes in hydraulic conductivity than other input parameters. From both scenario and sensitivity analyses, it has been found that some common hydrologic phenomenon prevails in the subsurface environment of the study area that leads to enrichment and migration of fluoride from the contaminated zone to the other part of the model domain along with groundwater flow. Although the calibrated model has a significant correlation coefficient for flow as well as for the transport model, however, the model has some limitations that are associated with quality and availability of accurate input parameters, assumption and simplification. For example, in the present model, uniform thickness of different layers has been assumed to avoid complexity, but actually, the stratigraphic layers are non-uniform.

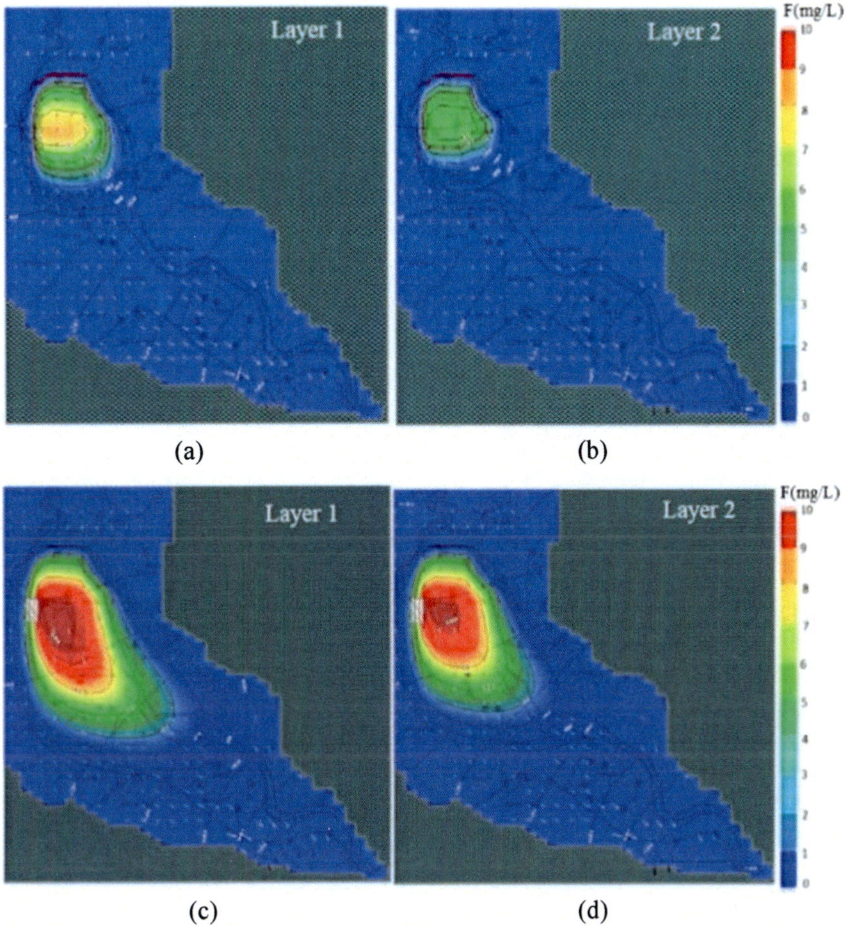

Fig. 5 Spatial distribution of fluoride plume after 10 years simulation period (**a**, **b**) and after 20 years simulation period (**c**, **d**)

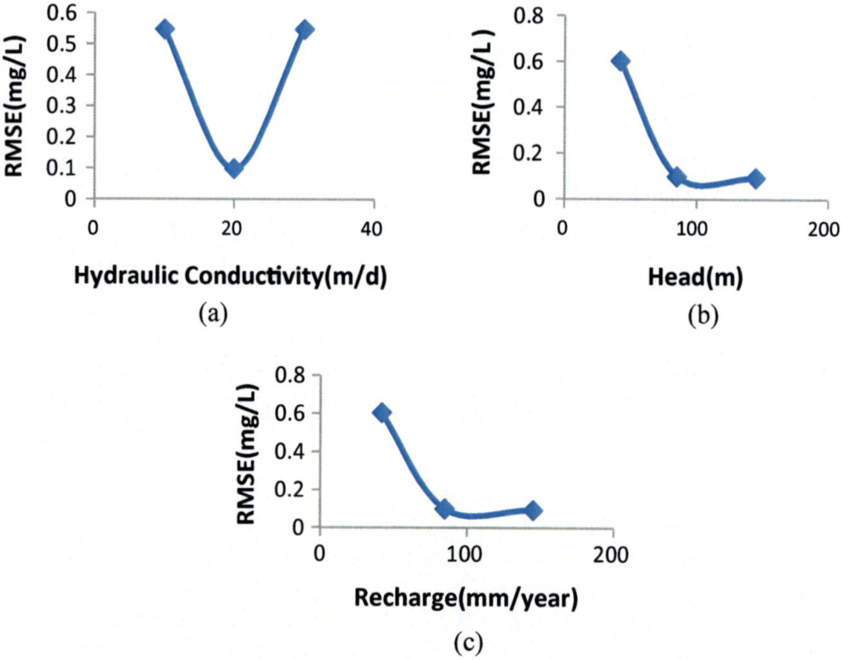

Fig. 6 Result of sensitivity analysis (**a**–**c**)

Acknowledgements. The authors would like to acknowledge West Bengal Public Health Engineering Department for providing necessary data to conduct this simulation study.

References

1. Amini M, Mueller K, Abbaspour KC, Rosenberg T, Afyuni M, Møller KN, Johnson CA (2008) Statistical modelling of global geogenic fluoride contamination in groundwaters. Environ Sci Technol 42(10):3662–3668
2. Ayoob S, Gupta AK (2006) Fluoride in drinking water: a review on the status and stress effects. Crit Rev Environ Sci Technol 36(6):433–487
3. Nagaraju A, Thejaswi A, Sun L (2016) Statistical analysis of high fluoride groundwater hydrochemistry in Southern India: quality assessment and Implications for source of fluoride. Environ Eng Sci 33(7):471–477
4. Ghosh A, Mondal S (2018) Application of multivariate statistics towards the geochemical evaluation of fluoride enrichment in groundwater at Shilabati river bank, West Bengal, India. Environ Eng Res 24(2):279–288
5. World Health Organization (2008) Guidelines for drinking-water quality: incorporating the first and second addenda. World Health Organization, Geneva
6. Andezhath SK, Ghosh G (2000) Fluorosis management in India: the impact due to networking between health and rural drinking water supply agencies. IAHS (International Association of Hydrological Sciences) Publi 260:159-165

7. WBPHED (2007) A note on fluoride contamination of groundwater in West Bengal. Public Health Engineering Department. Govt. of West Bengal, Kolkata. Available at www.wbphed.gov.in/flroride_cont.html
8. Zheng C, Wang PP (1999) MT3DMS: a modular three-dimensional multispecies transport model for simulation of advection, dispersion, and chemical reactions of contaminants in groundwater systems; documentation and user's guide. Alabama Univ University
9. Saghravani SR, Mustapha SAB, Ibrahim SB, Yusoff MK, Saghravani SF (2011) Phosphorus migration in an unconfined aquifer using MODFLOW and MT3DMS. J Environ Eng Landsc Manag 19(4):271–277
10. Ghoraba SM, Zyedan BA, Rashwan IMH (2013) Solute transport modeling of the groundwater for quaternary aquifer quality management in Middle Delta, Egypt. Alexandria Eng J 52(2):197–207
11. Reed EM, Wang D, Duranceau SJ (2017) Modeling anthropogenic boron in groundwater flow and discharge at Volusia Blue Spring (Florida, USA). Hydrogeol J 25(1):91–101
12. Freeze RA, Cherry JA (1979) Groundwater. Prentice-Hall, Englewood Cliffs, NJ, USA
13. Kent Anderson R (1967) Measurement of vertical and horizontal hydraulic conductivities on an undisturbed soil core. Electronic theses and dissertations, 3272
14. Davis SN (1969) Porosity and permeability of natural materials. Flow Through Porous Media 54–89
15. Xu M, Eckstein Y (1995) Use of weighted least-squares method in evaluation of the relationship between dispersivity and field scale. Groundwater 33(6):905–908
16. Ma R, Zheng C, Prommer H, Greskowiak J, Liu C, Zachara J, Rockhold M (2010) A field-scale reactive transport model for U (VI) migration influenced by coupled multirate mass transfer and surface complexation reactions. Water Resour Res 46(5)
17. Chandra S, Saksena RS (1975) Water balance study for estimation of ground water resources. Water Energy Int 32(4):443–450
18. GEC (Ground Water Estimation Committee) (1984) Report of the Ground Water Estimation Committee, Ministry of Irrigation, Government of India, New Delhi, 53
19. Ganguly S (2010) Evaluation of ground water potential of Silabati river basin West Bengal India. PhD theses
20. Zheng C, Bennett GD (2002) Applied contaminant transport modelling, vol 2. Wiley-Interscience, New York, p 353
21. Foglia L, Hill MC, Mehl SW, Burlando P (2009) Sensitivity analysis, calibration, and testing of a distributed hydrological model using error-based weighting and one objective function. Water Resour Res 45(6)
22. Ghosh A, Kartha SA, Mondal S, Dutta R (2019) Physical and numerical modeling of non-reactive solute transport in a heterogeneous laboratory aquifer model. Hydrol J (IAH) (42):27-40
23. Ting CS, Zhou Y, De Vries JJ, Simmers I (1998) Development of a preliminary ground water flow model for water resources management in the Pingtung Plain, Taiwan. Ground Water 36(1):20–36

Chapter 24
Sustainable Agriculture in a Cold Desert: Case Study of Lahaul and Spiti District of Himachal Pradesh

Anupama Shashni(✉) and Smita Sharma

Department of Economics, Panjab University, Chandigarh 160014, India

Abstract. Lahaul and Spiti district in Himachal Pradesh is known for its geographical ruggedness and harsh climatic conditions with dry summers and long freezing winters. The inhabitants of this region are largely dependent on agriculture for their livelihood. There is a dominance of small and marginal farms in Lahaul and Spiti due to the limited availability of cultivable land. Lahaul and Spiti has a dry temperate climate and receives negligible rainfall. It lies in the rain shadow area, so glacial water is the only source of irrigation. Inadequate snowfall during winter season and depletion of glaciers due to rise in temperature is posing a threat to the farmers who are completely dependent on these snow reservoirs for irrigation. This increases the risk of crop failure and poses a serious threat to the livelihood of farmers. There has been a shift in cropping pattern from traditional and resilient crops to economically rewarding crops like peas and cauliflower to get larger monetary benefits. The water-intensive nature of such crops is exerting pressure on the limited water resources in the region. The present study is based on a primary survey of two villages, 'Ghoshal' of Lahaul valley and 'Losar' of Spiti valley in the context of irrigation facilities. The survey results reveal that agricultural sustainability is highly dependent on the sustainability of water resources in the region.

Keywords: Irrigation · Sustainability · Agriculture · Cold desert · Lahaul and Spiti

1 Introduction

The scarcity of water resources arising due to climate change is posing a threat to the survival of farmers in the Himalayan region since they mostly rely on the traditional irrigation methods for farming. Glacier retreat generates serious concerns about water supplies for different communities living in and around the world's glacierized mountains [7]. Studies have found that the rise in the temperature at high elevations is already occurring at a threefold pace as compared to the global average. Such regions are very sensitive to the ongoing global climate change. Due to global warming, glaciers are getting depleted in the Himalayas posing a threat to the farmers who completely depend

R. K. Bhattacharjya et al. (eds.), *Sustainable Water Resources Management*,
Advances in Sustainability Science and Technology,
https://doi.org/10.1007/978-981-16-7535-5_25

on these snow reservoirs for irrigation. Glacial runoff due to fast melting of glaciers is reducing the water availability for irrigation, which further leads to the decline in the agricultural productivity and adversely impacts the livelihood sustainability of farmers [4].

The district Lahaul and Spiti of Himachal Pradesh represents a traditional agricultural society. It is known for its harsh cold climatic conditions and geographical inaccessibility. It is a cold desert region. Nearly, 80% of its population is engaged in agriculture and its allied activities. The district comprises of two major valleys, Lahaul valley and Spiti valley. There is a dominance of small and marginal farmers in Lahaul and Spiti due to the limited availability of cultivable land. The district falls in the rain shadow area of the Himalayas, so it receives negligible rainfall in the summers, and snowfall in the winters is the only source of water for irrigation and domestic usage. Due to rising temperatures, there has been a rapid reduction in the volume of glaciers leading to decreased water availability. Over the years, there has been a shift in cropping pattern from traditional and resilient crops to economically rewarding cash crops to get a larger monetary benefit. But the water-intensive nature of these cash crops makes the cultivation climatically unsustainable in the long run as water sources are drying up fast in this region. Thus, it may create a threat to the livelihood sustainability of people who are majorly dependent on agriculture for their livelihoods. The present study is based on the primary survey of two villages, 'Ghoshal' of Lahaul valley and 'Losar' of Spiti valley in context of irrigation facilities since the district is a cold desert area.

2 Methodology

Sampling

The present study is based on the primary survey of two villages, 'Ghoshal' of Lahaul valley and 'Losar' of Spiti valley. The primary data has been collected from 100 farm households. The selection of farm household has been done following multi-stage sampling procedure. In the first stage, both the blocks of district 'Lahaul and Spiti' were selected, namely Lahaul block and Spiti block, since both have different geographical and climatic conditions. In the second stage, one village was selected from each block after the consultation with the grass root functionaries and natives of the district, as it was found that not many villages were having agriculture as their main occupation. Village 'Ghoshal' was selected from the Lahaul block, and village 'Losar' was selected from the Spiti block. In the third stage, 50 farm households were selected from each village through convenient sampling method. The large farmers were excluded from the study since the majority of farmers in the district were marginal, small or medium farmers. The data from sample households was collected through a personal interview method in the year 2018. The secondary data was collected for twelve years from 2005–06 to 2016–17 due to availability of meaningful data from reliable sources for this period only. The secondary data has been collected from various issues of Statistical Abstract of Himachal Pradesh and the records of state government departments.

Compound Annual Growth Rate (CAGR)

To find the growth rate of variable Y over time t, we write the formula of compound interest rate, $Y_t = Y_0(1 + r)^t$, where Y_t = value of variable Y at time t and Y_0 = initial value of the variable. When equation $Y_t = Y_0(1 + r)^t$ is subjected to logarithmic reductions, we get $l_nY_t = l_nY_0 + tl_n(1 + r) + u_t$, where u_t is the disturbance term. By substituting $\beta_1 = l_nY_0$ and $\beta_2 = l_n(1 + r)$ in the above equation, we get,

$$l_nY_t = \beta_1 + \beta_2 t + u_t \tag{1}$$

where β_1 and β_2 are parameters of the linear regression model, and β_2 gives the instantaneous rate of growth. Compound Annual Growth Rate (r) can then be evaluated by taking the antilog of estimated β_2 and subtracting 1 from it and multiplying the difference by 100, i.e.

$$r = [\text{antilog}(\beta_2) - 1] \times 100 \tag{2}$$

The model in Eq. (1) is a linear regression model where regressand is the logarithm of Y, and the regressor is time. From this linear regression model, we get the value of β_2 which gives the instantaneous rate of growth, which is substituted in Eq. (2) to get the Compound Annual Growth Rate.

Herfindahl Index (HI)

Herfindahl Index (HI) is defined as the sum of the square of all n proportions is a measure of crop concentration. This measure is used to measure crop diversification on acreage proportion.

$$HI = \sum_{(i=1)}^{n} P_i^2$$

where $P_i = Ai/\sum_{i=1}^{n} A_i$; P_i = proportion of ith crop, A_i = area under ith crop (in hectares), ΣA_i = total cropped area (in hectares), i = 1,2,3,4,…n (number of crops).

Percentages

Simple percentages were used to analyse the results of the primary survey conducted in village Ghoshal of Lahaul valley and village Losar of Spiti valley. The main motive was to explore different aspects of cropping pattern and irrigation system in the selected district.

3 Review of Literature

Several studies have stated that the Himalayan region is very sensitive to the ongoing climate change due to its fragile ecosystem. Ladakh and Lahaul and Spit region of

Himalayas have witnessed a visible rise in the temperature and decline in precipitation leading to less availability of water in the glacial streams [1]. The climate change has also impacted the agriculture in the form of reduced water availability for irrigation, shifts in rainfall, drought conditions and a decline in the overall crop yield [4, 8]. Studies have highlighted the role of social institutions and social capital to deal with the adverse impacts of climate change in Lahaul and Spiti district of Himachal Pradesh [9].

4 Cropping Pattern and Present Irrigation System in Lahaul and Spiti

The results of the field survey as well as secondary data analysis have been discussed in this section. The field survey was conducted in village Ghoshal of Lahaul valley in June 2018 and village Losar of Spiti valley in September 2018. Fifty farm households were selected from each village for analysing different aspects of cropping pattern and irrigation system in the selected district.

But before conducting the field survey, an attempt was made to analyse the cropping pattern of Lahaul and Spiti district using secondary data. So, the Compound Annual Growth Rate (CAGR) and Herfindahl Index (HI) were calculated using the secondary data. To calculate the Compound Annual Growth Rate and Herfindahl Index for district Lahaul and Spiti, we have used seven crops, which is a mix of food grains and horticultural crops like wheat, barley, potato, green pea, cauliflower, cabbage and apple. The time period taken for the study is 12 years starting from 2005–06 to 2016–17.

It can be seen from Table 1 that there was a significant increase in area, production and yield of cauliflower and cabbage, whereas there was a significant decline in area, production and productivity of green pea during the study period. During the primary survey, it was found that the increase in area under cauliflower was mainly due to its better returns as compared to green pea. There was also a positive growth rate in the area under apple plantation in the district during the study period. This was mainly due to climate change due to which areas suitable for apple cultivation have shifted to higher altitude region like Lahaul and Spiti. In addition to that, it was also found that apple grown in Lahaul fetches a good price in the market, so many farmers have started apple plantation to reap its economic benefits.

Table 2 shows Herfindahl Index values during the study period, i.e. from 2005–06 to 2016–17. There was an overall decline in the HI values during the study period from 0.45 to 0.29 implying an increase in crop diversification in Lahaul and Spiti. When the value of the Herfindahl Index (HI) exceeds 0.5 in a particular region calculated for a given set of crops, then it is moving towards specialization. There was a sharp decrease in the value of Herfindahl Index to 0.32 in the year 2014–15 implying a sharp increase in diversification in the region. This was mainly due to a sudden increase in area under cauliflower cultivation from 67 to 607 hectares between the years 2013–14 to 2014–15. From the year 2014–15 onwards till the year 2016–17, there was a gradual decline in the values of Herfindahl Index from 0.32 to 0.29 implying an increase in the level of crop diversification. This was mainly due to a diversion of land from pea cultivation to cauliflower cultivation. After interviewing the farmers, it was found that diversification towards new cash crops like cauliflower was due to improved availability of modern

Table 1 Compound annual growth rate (CAGR) of area, production and productivity for major crops in Lahaul and Spiti (2005–06 to 2016–17)

Crop	Area	Production	Yield
Cauliflower	35.25%***	43.90%***	6.39%***
Potato	−1.30%N	−2.42%N	−1.10%N
Green pea	−1.61% **	−2.73%***	−1.10%**
Cabbage	6.07% *	15.02%***	8.32%***
Wheat	2.32%N	5.86%***	8.32%***
Barley	−1.10%N	9.08%***	10.29%***
Apple	8.49%***	−0.60%N	−11.40%**

Source Calculated by the author
***—significant at 1%, **—significant at 5%, *—significant at 10%

inputs required for the cultivation of new cash crops, extension services provided by government agencies, good prices for new horticultural crops and climate change.

Table 2 Values of Herfindahl index from 2005–06 to 2016–17

Year	Herfindahl index
2005–06	0.45
2006–07	0.44
2007–08	0.43
2008–09	0.45
2009–10	0.44
2010–11	0.41
2011–12	0.40
2012–13	0.40
2013–14	0.41
2014–15	0.32
2015–16	0.31
2016–17	0.29

Source Calculated by the author

The field survey results revealed that lack of proper irrigation facility is one of the major constraints since district Lahaul and Spiti receives negligible rainfall during the cultivation period, and the whole crop is dependent on the limited snow meltwater from glaciers which is used for irrigation through Kuhls, i.e. the water channels that are used for irrigating the fields. Around 60% of farmers interviewed in village Ghoshal and 90% of farmers in the village Losar said that lack of proper irrigation facility is a major constraint faced by them. The farmers said that they have to contribute labour to construct and maintain the Kuhls. The cemented Kuhls constructed by the government agencies are of poor quality and lead to water wastage if not repaired every year since they get damaged due to snow frost during the winters. If they are not repaired timely, it leads to soil erosion and wastage of limited water resource. Hundred per cent farmers interviewed in Losar village demanded good quality cemented Kuhls which are timely repaired every year before the cropping season begins to check any kind of water wastage and to ensure timely availability of water for irrigating the fields. Sprinklers provided by the government agencies at subsidized prices are also used for irrigating the fields, but Kuhls are majorly used for irrigation. The efficient utilization of water resources is very important for increasing agricultural productivity and hence increasing the income of the farmers.

5 Kuhl as a Common Property Resource

Common property resources are those natural resources which a community or society owns and manages collectively. Since the limited natural resources can be consumed to the point of overexploitation, they are considered to be rivalrous. In addition to that, exclusion from accessing common property resource is difficult, so it is also considered as non-excludable. The Kuhls have a property of being quasi-public good, which in the wake of depleting water resources has become non-excludable and rival. The Kuhls can be treated as a common property resource since they are managed by the Kuhl committee (Water institution) which takes all the important decisions regarding the construction and maintenance of the Kuhls and also regarding the distribution of water rights. These Kuhl committees are created at the village level to regulate the usage of glacial water. The Kuhls are small gravity channels constructed along mountainsides to harness water from the glacial streams to the agricultural fields [5]. The water from snow-fed nullahs is carried through channels to the desired place [2]. Irrigation is mainly done by an elaborate Kuhl system in Lahaul and Spiti which uses glacial water for irrigation. Similar Kuhl regime was also found in Kangra district of Himachal Pradesh for the management of water resources for irrigation through Kuhls [3].

The distribution of water rights to the farmers is done by using Bari system (turn System) since the availability of water resources is limited in the district. Each household has to contribute labour for the construction and maintenance of the Kuhls. Kuhls are built by collective labour of the village, and the water is distributed according to the share of land owned by a family. It leads to equal distribution of limited natural resources. The distribution of water rights is done through a lottery system by which sequence of water rights is decided. The motive is to use limited water resources most efficiently. Failure to contribute family labour leads to penalties and even withdrawal of water rights. Each farmer has to wait for his turn to irrigate and cannot afford to miss even a single turn because missing even one turn may cause crop failure since they may sometimes have to wait for several days for their turn to come. So, this time constraint for irrigating the fields becomes a cause of stress among the farmers.

It was also found during the primary survey that water rights distribution rules were less stringent in areas where water for irrigation is in plenty. On the other hand, water distribution rules were made strict in water-scarce villages to avoid inequity and fights over water distribution. This distinction in strictness of rules was visible in the two villages surveyed for the present study. It was found that Village Ghoshal had comparatively less stringent water distribution rules as compared to village Losar due to lesser water scarcity. Water availability has been a serious problem in the district due to climate change and dependence on glacier meltwater for irrigation and domestic purposes. Spiti region faced a critical water scarcity problem in 2004–05. All the natural springs had dried up and not even sufficient water for drinking was left in the upper lying villages in Spiti like Langsa, Kibber and Dimul. As a result, few families in Langsa disbanded everything and migrated to other villages in Spiti [1, 9]. So, it becomes important to regulate the limited water resource through local level common property protocols in this cold desert region. It enables the communities to create their own rules to prevent overuse and ensure equal access to water resources for all the farmers. Such water management institutions reveal the capability of the communities to deal with various kinds of shocks and uncertainties.

6 Conclusion

The district Lahaul and Spiti of Himachal Pradesh represents a traditional agricultural society with nearly 80% of the population engaged in agriculture and its allied activities. The survey results reveal that agricultural sustainability is highly dependent on the sustainability of water resources in the region. Over the last few years, impacts of global warming have been visible in Lahaul and Spiti district in the form of the reduced amount of water in the glacier-fed streams [1]. With increasing uncertainty, climate change and water scarcity, it becomes essential to adopt certain rules and regulations at the community level to sustainably maintain the scarce water resources which will ultimately help in sustaining farmers livelihoods [9]. The Kuhls as a common property resource being managed by the Kuhl committee (water institution) have helped in regulating the usage of glacial water efficiently. These Kuhl committees constituted in various villages can be seen as playing an adaptive role to deal with the problem of scarcity of water resource in the district in the wake of ongoing climate change. The shift in cropping pattern towards more water-intensive cash crops has also been exerting pressure on limited water resources in the region. In such a situation, it becomes important to research on water-saving irrigation methods in Lahaul and Spiti, keeping the seasonal damage and infrastructural costs in mind. There is also a need to introduce new varieties of crops with greater tolerance towards drought conditions. Timely repair of cemented Kuhls and tanks should also be done to avoid wastage of water through the leakage. Policymakers should also check the feasibility of constructing innovative water harvesting structures like ‘Artificial Glacier’ and ‘Ice Stupas’ in Lahaul and Spiti which have been built in Ladakh since both the regions have similar geo-climatic conditions. These two structures work on different mechanisms but both aim at storing the glacial water in frozen form during the winter months and discharging the water to the fields during the sowing season in March and April when water is scarce for irrigation [6, 10]. All such measures can help in mitigating the negative effects of water scarcity in the environmentally fragile region of Lahaul and Spiti.

References

1. Angmo T, Mishra SN (2009) Impact of climate change in Ladakh and Lahaul and Spiti of the western Himalayan region. In: Proceedings of GERES seminar on Energy and climate change in cold regions of Asia, pp 51–54
2. Bajpai SC (1987) Lahaul-spiti: a forbidden land in the Himalayas. Indus Publishing Company
3. Baker J (1999) Persistence, transformation and demise within the gravity flow irrigation systems (Kuhls) of Kangra Valley in Himachal Pradesh. In: Workshop on the cooperative management of water resources in South Asia. Indiana University
4. Carey M, Molden OC, Rasmussen MB, Jackson M, Nolin AW, Mark BG (2016) Impacts of glacier recession and declining meltwater on mountain societies. Ann Am Assoc Geogr 1–10
5. Chand R (1996) Ecological and economic impact of horticultural development in the Himalayas: evidence from Himachal Pradesh. Econ Polit Weekly 93–99
6. Clouse C (2014) Learning from artificial glaciers in the Himalaya: design for climate change through low-tech infrastructural devices. J Landscape Archit 6–19
7. Intergovernmental Panel on Climate change (2014) Climate change 2014: impacts, adaptation and vulnerability. Part A: global and sectoral aspects. Cambridge University Press, New York
8. Negi GCS, Samal PK, Kuniyal JC, Kothyari BP, Sharma RK, Dhyani PP (2012) Impact of climate change on the western Himalayan mountain ecosystems: an overview. Int Soc Trop Ecol 345–356
9. Padigala B (2015) Social capital and local institutions: a perspective to assess communities adaptation potential to climate change. Handbook of Climate Adaptation, pp 1–21
10. Shaheen FA (2016) The art of glacier grafting: innovative eater harvesting techniques in Ladakh. In: International Water Management Institute-TATA Water Policy Program, pp 1–6